Contents

Foreword   Murray Gerstenhaber   v

The Ecological Role of Volterra's Equations   Egbert R. Leigh   1

Evolution of Complex Genetic Systems   Richard C. Lewontin   62

Studies on Biological Clocks: A Model for the Circadian Rhythms of Nocturnal Organisms   Theodosios Pavlidis   88

Author Index   113

Subject Index   115

Foreword

Biologists have become increasingly sophisticated in their use of mathematics, but there has been little direct contact between them and mathematicians. In the hope that more direct communication would be useful to both, the American Mathematical Society and the Society for Industrial and Applied Mathematics have jointly undertaken the sponsorship of a series of annual symposia entitled "Some Mathematical Questions in Biology". This volume contains the proceedings of the first symposium, held December 28, 1966, in Washington, D.C., during the 133rd annual meeting of the American Association for the Advancement of Science.

Modern biology has long passed the purely descriptive phase, and for many phenomena, biologists have now developed elegant mathematical models. The main contribution of mathematics to biology, as to any science, has been the concepts out of which the models are built. But it has been characteristic of biology that some of the most important advances have come through the use of relatively simple mathematical concepts, like those of stability and of a code. In every science which is becoming "mathematical", there is the danger of considering simple things unimportant, and of trying prematurely to apply refined mathematical theories and powerful computational tools just because one has them at hand. This danger is even more acute in biology, and the main point when biologists and mathematicians meet is to keep the biological questions clear. For this reason, most speakers at these symposia have been biologists, and all have tried to educate the listening mathematician to the basic biological questions involved. In every case some mathematical question arose, but the attempt has been to keep the mathematics relevant to the biology, rather than to let it be an end in itself.

It is a pleasant duty to record my thanks to Professors Robert MacArthur and Lionel Rebhun of the Department of Biology, Princeton University, for help in organizing the symposium, and to the United States Steel Foundation for financial support.

          Murray Gerstenhaber
          Department of Mathematics,
           UNIVERSITY OF PENNSYLVANIA
          July 10, 1968

The Ecological Role of Volterra's Equations

Egbert G. Leigh, Jr

Abstract and acknowledgements. I will open with a historical sketch of ecologic theory. This is mostly to provide a perspective for what follows. I have focussed on the extrapolation of theory from simple systems capable of a rather full mathematical analysis to more complex natural systems. The ecologist, like the student of statistical mechanics, has had to settle for a rather grosser characterisation of a community than perhaps he first expected: finding a suitable characterisation is, in a sense, his major problem.

Then I shall outline a "statistical mechanics of biological associations" first developed by Kerner. This may be used to calculate the nature and strength of the interactions between two species from knowledge of their population fluctuations. The theory may also be used to relate general features of community organisation to each other.

Next follows a description of the relation of this "statistical mechanics" to certain Gaussian Markov processes describing the behavior of communities in fluctuating environments.

Finally I shall use the techniques developed for the purpose to discuss the possible mechanisms for an oscillation of lynx and hare populations in Canada. I shall only succeed in showing that it is not a predator-prey oscillation.

I am indebted to many people for being able to write this paper. To the N.A.S.A., which supported me quite magnificently as a graduate student, when I did the greatest part of the work described here; to Dr. MacArthur, for many valuable discussions, and for introducing me to the intricacies of two successive desk calculators, which allowed me so handily to develop the covariances; to the ghost of Haydn, whose music makes writing mathematics tolerable. I am also more deeply grateful than I can say to Dr. Gerstenhaber for his patience with a very slow author.

0. **Introduction.** How does one construct a meaningful mathematical theory of ecology? To do so we must decide what we wish to talk about and how we wish to talk about it. A great many of these decisions were made by the pioneers of mathematical ecology, Lotka [1] and Volterra [2], so it is well to begin by reviewing their approaches.

Both Lotka and Volterra focussed on the "biological association", a group of species each of which is influenced by the others in one way or another, but none of which are influenced by species outside the association. Such closed associations do not occur in nature. However, when an ecologist reads of a biological association he thinks of a community, comprising the association of plants and animals living together in a particular setting, such as a lake, a coral reef, or an expanse of virgin forest. And surely he is not far wrong in identifying the ecologist's community concept with the association of Lotka and Volterra.

How does one characterise a community? Lotka and Volterra characterised it by a state-vector having as many components as there are species in the community: the ith component of this state-vector, N_i, is simply the population size of the ith species in the community. These population sizes are usually represented as population densities, so that N_i is the number (or weight) of species i per unit area. The reasoning of ecologic theory has been developed in more or less conscious imitation of chemical kinetics, in which concentrations (rather than absolute amounts) play a major role, so it is natural for the ecologist to focus on population densities.

How does the state-vector change with time? More specifically, what governs the change in the population density N_i of species i? We may suppose that if species i found itself all alone in the setting of our community, its numbers would change by geometric progression. It would increase in numbers if it were a primary producer, an organism, like a plant, which does not depend on other organisms for its food, and it would decrease otherwise. Thus we write
$$dN_i/dt = e_i N_i.$$
We suppose, moreover, that the effect of species s on the growth rate dN_i/dt of species i is proportional to the number of meetings between them, which we assume to be proportional to the product $N_i N_s$ of their population sizes. Thus we write
$$\frac{dN_i}{dt} = e_i N_i + \sum_{s=1}^{n} a_{is} N_i N_s$$
where the summation represents the effects of interactions with all the other species in the community on the growth rate of species i. Notice that if species i preys on species s, a_{is} will be positive and a_{si} negative, since species i is benefiting at the expense of species s. If, on the other hand, these species are competing for room to grow, like barnacles and mussels on a crowded rock or tall trees reaching for the light, a_{is} and a_{si} will both be negative, since the growth rate of each species is lowered to some extent by the presence of the other. By such arguments Volterra arrived at this "dynamical law" for communities.

Lotka came by this law in a more abstract fashion. He reasoned from general principles that
$$dN_i/dt = F_i(N_1, \ldots, N_n).$$

If a species is absent from a closed community, it stays absent. Thus $N_i = 0$ implies $F_i(N_1, \ldots, N_n) = 0$. To incorporate this fact, Lotka wrote

$$F_i(N_1, \ldots, N_n) = N_i G_i(N_1, \ldots, N_n).$$

The simplest feasible expression for G_i is

$$G_i(N_1, \ldots, N_n) = e_i + a_{i1} N_1 + \cdots + a_{is} N_s.$$

Following in the footsteps of their physicist predecessors, Lotka, Volterra and their contemporaries sought to describe the behavior of simple communities that could be created in the laboratory. Thus for a single species with a limited food supply, their theory yields the equation of growth

$$dN/dt = eN - aN^2.$$

If we set $e/a = K$, this equation becomes

$$dN/dt = aN(K - N)$$

from which we learn that the population N increases geometrically when very rare, but gradually levels off toward the equilibrium value K, the maximum population density it can maintain.

Lotka and Volterra discussed a number of two-species systems, perhaps the most notable case of which was the competition of two species for the same space. The appropriate equations for the populations of two competing species are

$$dN_1/dt = e_1 N_1 - a_{12} N_1 N_2 - a_{11} N_1^2,$$
$$dN_2/dt = e_2 N_2 - a_{21} N_2 N_1 - a_{22} N_2^2.$$

Perhaps the most interesting experimental example [3] is that of two yeasts competing in an anaerobic medium.

Yeasts, of course, make alcohol, but their growth rate is the lower the higher the alcohol content of their medium. If we measure the populations of the two yeasts by their contributions to the alcohol content of the medium (hoping that, for each yeast, population size is proportional to alcohol concentration) we obtain

$$dN_1/dt = e_1 N_1 - a_1 N_1 (N_1 + N_2),$$
$$dN_2/dt = e_2 N_2 - a_2 N_2 (N_1 + N_2)$$

where $N_1 + N_2$ is the alcohol content of the medium. If we set $e_1/a_1 = K_1$, $e_2/a_2 = K_2$, we find

$$dN_1/dt = a_1 N_1 [K_1 - (N_1 + N_2)],$$
$$dN_2/dt = a_2 N_2 [K_2 - (N_1 + N_2)].$$

K_1 is the alcohol concentration at which species 1 can just maintain itself, and K_2 is the same for species 2. Let $a_2/a_1 = x$. Then we may write

$$\frac{d}{dt} \log \frac{N_1}{N_2^x} = \frac{d}{dt} \log N_1 - x \frac{d}{dt} \log N_2 = a_1(K_1 - K_2).$$

Thus only one of the two species will persist in this system, that with the higher alcohol tolerance. This is the simplest example of the "principle of competitive exclusion."

Also famous was the system consisting of one predator species and its prey. Letting N_1 be the prey population and N_2 the predator, we may write

$$dN_1/dt = e_1 N_1 - a_1 N_1 N_2 = a_1 N_1 (K - N_2),$$
$$dN_2/dt = -e_2 N_2 + a_2 N_1 N_2 = a_2 N_2 (N_1 - R).$$

These equations represent the fact that the prey population increases in the predator's absence and is harmed by its presence, while the predators decrease in the absence of prey and benefit from their presence. In particular, the prey will increase unless the predators exceed a certain maximum tolerable abundance K, while the predators require a certain minimum abundance R of the prey in order to increase. One consequence of these equations, as we shall show later, is that the populations of predator and prey will both exhibit cyclic fluctuations of abundance. Another curious result is that if both predator and prey are harvested at a rate proportional to their abundance (thus increasing e_2 and decreasing e_1), the mean population of the predators will decrease while that of the prey will increase. It is self-defeating to employ an insecticide on an insect already controlled by a natural predator.

These developments greatly stimulated ecological thinking. The principle of competitive exclusion, which states that if two species in a community live the same way of life, only one will survive, is pregnant with consequences [4]. By this principle we understand how each species in a community must be characterised by a distinct way of life. We understand also the phenomenon of "character displacement": two related species which overlap over part of their range will be less similar where they occur together than elsewhere, for where they overlap they must keep out of each other's way. More importantly, we gain some insight into what gives the advantage to one species over another in a competition. An analysis of the dynamics of yeast competition shows that, in an anaerobic medium, the two species affect each other through their alcohol production and tells us that the species with the greater alcohol tolerance will win out. Similarly, trees competing for light affect each other by the leaf area they expose to the sun: the tree which can maintain the greater leaf area per unit ground area wins out. An analysis of two predators competing for the same prey tells us that the predator requiring the lower prey abundance for his maintenance wins out, while if two prey are limited by the same predator, that species which can support the larger abundance of predators wins out. On such a basis one can elaborate a considerable body of semiqualitative theory concerning the allocation of related roles in a community, when a generalist feeding on several foods is favored over specialists on each, and so on [5]. We come to recognise a basic difference between the dynamics of space-limited species, on the one hand, and those of species limited by relationships of predator and prey (food web relationships) on the other: the one characterised by a speedy approach to a steady state, the other by continuing oscillations. Some think plants [6] are primarily limited by sunlight and water, rather than by the animals which eat them: their competition is mathematically a form of competition for space. Animals, on the other hand, are usually limited by food web relationships. Thus the mathematical distinction corresponds to a basic biological one.

The original motivation for studying simple experimental communities was, however, rather different. People hoped thereby to develop and refine the theory to an extent sufficient to permit direct extrapolation to more complex natural communities. This program was a failure, for the curious reason that simple experimental communities are too unstable and erratic to exhibit lawful behavior. The population of the United States from 1790 to 1940 [7] exhibits a far closer fit to the logistic growth-law than most experimental populations (although the second World War ruined the fit by seemingly greatly increasing the population this country could support). Population oscillations predicted by the simple theory of Volterra and Lotka occur only in the most complex (and most nearly natural) laboratory situations: usually the predators eat up all the prey and then die out themselves. In competition for space, it is difficult to evaluate the interaction coefficients accurately: the theory is to this extent imprecise in its predictions. And, in general, it appears that the more complex an ecological system, the more orderly will be its behavior.

Thus it appears that a detailed solution of the dynamical equation of a community would be biologically useless even if it were mathematically feasible. If this is so, then what questions, if any, should we ask of this theory, and how may we go about answering them? Quite typically, we have found a means for answering our questions before discovering what questions to ask. The means in question is inspired by the example of statistical mechanics. It is a familiar story how the perfect gas law can be deduced from the crudest picture of molecular billiard-balls bouncing hither and yon in a box. Somehow, in the attempt to relate gross features of the gas to each other, the errors in our description of the motions of the individual molecules cancel out. Khinchin's [8] strange and suggestive work on statistical mechanics presents the subject in the form of a statistical theory of differential equations. The fundamental aim of this theory is to derive a joint probability distribution for the state-vector of the system: having done this, we may calculate the average of any quantity of interest to us. The derivation of this distribution strikingly parallels that of the central limit theorem, and merits discussion in its own right. It so happens that the dynamical equations of animal communities are of a type to which the theory applies: this fact was discovered and put to brilliant use by E. H. Kerner [9] in a series of reports to the Bulletin of Mathematical Biophysics. It is this statistical mechanics of population dynamics which I wish to explore in this paper.

What parameters of a community is it suitable to try to relate [10]? For a long time people have been concerned with the productivity of a field under different types of agriculture: what return in food value does one obtain from the seed put into the ground? One might ask the more abstract question: if a field is farmed in a certain way, how much food (in calories per year) will it supply its farmer? Similarly, one might measure the productivity of a natural community by the sum of the feeding rates of all the predators in the community. This sum is also the rate at which food is supplied to them. If most of the organisms in the community are eaten, this sum will indeed be very nearly the rate of production of living matter in the community, the community *productivity*. The total weight of living matter in the community, the community *biomass* or standing crop, is another variable of importance. We shall also be concerned with the stability of a community. What could we mean by stability? In some communities the component species exhibit rather violent population crashes and explosions: lemmings [14] provide the best-known example of such a species, but insect plagues and the like provide others. A violent population explosion in one species threatens the balance of the community as a whole: indeed, the human population explosion threatens the balance of the whole earth. The stability of an individual species may be measured by the frequency of crashes or explosions in its numbers: we shall define the stability of the community as a whole to be some appropriate average of the stability of its component species. A large part of this report will be devoted to deriving a relationship between such macroscopic community-variables.

We shall be concerned also with the appropriateness of the statistical formalism. Its application requires that we be able to measure population sizes in such a way that the loss in prey population to a particular predator is precisely equal to the predator's gain. (In terms of the dynamical theory, the requirement is that $a_{is} = -a_{si}$.) This assumption is needed to establish the analogue of the energy conservation law on which the statistical formalism hinges. If we measure populations by weight, this relationship does not hold: it takes ten pounds of feed to make a pound of beef. We shall rectify the problem by measuring predator populations in terms of their prey equivalents, measuring an herbivore population in terms of the grass eaten to grow it, and a carnivore population in terms of the grass eaten by the herbivores the carnivores eat, etc. We find it difficult to measure an omnivorous population. Humans eat both meat and grain: is the missionary the cannibals eat worth ten or a hundred times his weight in grass? We therefore restrict ourselves to communities stratified into trophic levels such that the species of one trophic level feed only on those of the level below. We should also remember that the statistical theory treats populations interacting according to a "conservative" dynamics. A population curve $N_i(t)$ is an almost-periodic function whose amplitude of fluctuation depends on initial conditions: this amplitude is thus in some sense random. On the other hand, intuition tells the biologist that the amplitude of population fluctuations represents some sort of balance which is restored when disturbed. The most likely mechanism for such a balance is the interaction between environmental disturbance and influence tending to damp out population fluctuations. The damping influence might arise from "dissipative" interactions, such as a tendency for population increase to depress population growth rate (a_{ii} negative). In the latter part of the paper we shall explore a Gaussian Markov process representing a balance between damping tendencies and environmental disturbance.

The correspondence between the statistical dynamics of conservative systems and that of Gaussian Markov processes is very curious and merits some notice in its own right. I rather suspect this correspondence may provide grounds for yet another "natural" characterisation of the Gaussian distribution. This distribution has a curious habit of cropping up in all sorts of dynamical theories. An imaginary version (e^{-ix^2}) [11] plays the same role in Feynman's path integral theory of quantum mechanics that the real version does in Brownian motion formalisms. A series of strange and stimulating researches by Dr. Edward Nelson [12] of Princeton University tell us how, at least under some conditions, Schrödinger's equation of quantum mechanics can be derived by assuming a particle moves according to the law

$$dx(t) = b[x(t), t]\, dt + dB(t)$$

where x is the position of the particle and $dB(t)$ is a Langevin force term. In effect, Nelson has assumed the particle is subject to an elementary Brownian motion. His demonstration is somewhat involved, and he never gives an explicit expression for $b[x(t), t]$ in terms of familiar quantities such as the Hamiltonian. Yet surely the time will come when quantum mechanics will find a more natural expression in terms of Brownian motion. Gibbsian statistical mechanics in some ways foreshadowed quantum mechanics. From a linguistic point of view it is appropriate to ask why Gaussian Markov formalisms should play the role they do in this half-world between mechanics and quantum mechanics. One answer would be a formulation bringing out the points in common between these theories and making simple and natural the passage from one to the other. Such an answer would very much interest a biologist who well remembers Wiener's [13] conviction of the importance of Gaussian Markov processes in the future mathematics of "complex systems".

REFERENCES: INTRODUCTION

1. A. J. Lotka, *Elements of mathematical biology*, Dover, New York, 1956.
2. V. Volterra, *Lecons sur la theorie mathematique de lutte pour la vie*, Gauthier-Villars, Paris, 1931.

Save only Ronald Fisher, Volterra was the most gifted mathematician to devote his attention to biology during this period. It is curious the extent to which a sense of elegance and proportion, a sense of what to ask, serves in place of biological knowledge. It is the lesser talents, not the greats of the field, who have created the stock character of "the mathematician in biology".

3. This experiment, and many other competition experiments, are reported in
G. F. Gause, *The Struggle for existence*, Hafner, New York, 1934.
A corresponding array of predator-prey experiments are described in
G. F. Gause, *Verifications experimentales de la theorie mathematique de la lutte pour la vie*, Hermann et Cie, Paris, 1935.
See also
L. B. Slobodkin, *Growth and regulation of animal populations*, Holt, Rinehart and Winston, New York, 1962.

4. The uses of the principle of competitive exclusion are best set forth in

G. E. Hutchinson, *The ecological theater and the evolutionary play*, Yale Univ. Press, New Haven, Conn., 1965.

A rather briefer account is to be found in a paper, perhaps the best short summary of what ecology is all about, and a marvel for its beauty:

G. E. Hutchinson, *Homage of Santa Rosalia, or why are there so many kinds of animals?*, Amer. Naturalist **93** (1959), 145–159.

5. Some idea of the scope of such theory may be gleaned from

R. H. MacArthur, *Patterns of species diversity*, Bio. Rev. **40** (1965), 510–533.

6. Hairston, Smith and Slobodkin, *Community structure, population control, and competition*, Amer. Naturalist **94** (1960), 421–425.

The idea is quite controversial, partly because ecologists do not really believe a short paper can be significant, but partly because we do not really know enough to criticise it.

7. The best way to judge the fit is to check the applicability of the equation $d \log N/dt = e - aN$: simply plot $\Delta N/N$ against N, where ΔN is the change in population between successive U.S. censuses and N is the harmonic mean of the censuses being differenced to obtain ΔN.

8. A. I. Khinchin, *Mathematical foundations of statistical mechanics*, Dover, New York, 1949.

9. E. H. Kerner, *A statistical mechanics of interacting biological species*, Bull. Math. Biophysics **19** (1957), 121–146.

E. H. Kerner, *Further considerations on the statistical mechanics of biological associations*, Bull. Math. Biophysics **21** (1959), 217–255.

10. A famous discussion of this problem is

R. L. Lindeman, *The trophic-dynamic aspect of ecology*, Ecology **23** (1942), 399–418.

11. M. Kac, *Probability and related topics in physical sciences*, Interscience, New York, 1958.

12. E. Nelson, *Dynamical theories of Brownian motion*, Princeton Univ. Press, Princeton, N.J., 1967.

13. N. Wiener, *Nonlinear problems in random theory*, Wiley, New York, and M.I.T. Press, Cambridge, Mass., 1958.

14. C. Elton, *Voles, mice, and lemmings*, Oxford Univ. Press, New York, 1942.

I. THE STATISTICAL THEORY OF VOLTERRA'S EQUATIONS

1.1.1 We begin with a study of elementary interactions. As we are concerned with animal communities, we shall assume predation is the only basic interaction. The simplest possible system thus consists of a single predator species and his prey. Letting N_1 be the prey population size and N_2 that of the predator, we have

$$dN_1/dt = e_1 N_1 - a_{12} N_1 N_2,$$
$$dN_2/dt = -e_2 N_2 + a_{12} N_1 N_2.$$

Here we measure predator populations in terms of prey equivalents. If it takes ten pounds of antelope to make a pound of lion, we shall treat a pound of lion as ten pounds of antelope. We may write

$$(d \log N_1)/dt = e_1 - a_{12} N_2 = a_{12}(K - N_2),$$
$$(d \log N_2)/dt = -e_2 + a_{12} N_1 = -a_{12}(R - N_1).$$

Notice that

$$\frac{d \log N_1}{dt}(R - N_1) + \frac{d \log N_2}{dt}(K - N_2) = 0,$$

$$R \log N_1 - N_1 + K \log N_2 - N_2 = \text{const}.$$

If we plot the second equation on a graph of N_1 against N_2 we find it describes a closed curve: we may therefore conclude the populations exhibit cyclic fluctuations. The system is neutrally stable.

1.1.1.5 We may integrate the equation

$$\frac{d \log N_1}{dt} = a_{12}(K - N_2)$$

to obtain

$$\log N_1(t) - \log N_1(0) = a_{12}\left[Kt - \int_0^t N_2(t)\,dt\right].$$

We find

$$\lim_{t\to\infty} \frac{1}{t}\int_0^t N_2(t)\,dt = K - \lim_{t\to\infty} \frac{1}{a_{12}t}[\log N_1(t) - \log N_1(0)].$$

Since population sizes are executing cyclic fluctuations of some fixed amplitude, there must be some constant c such that, for all t,

$$|\log N_1(t) - \log N_1(0)| < c.$$

Thus the average population size of species 2,

$$\lim_{T\to\infty} \frac{1}{T}\int_0^T N_2(t)\,dt = K.$$

Similarly, the average population size of species 1 is R. The average population sizes of the two species are just those which would suffice to maintain each other's populations in equilibrium.

1.1.2 We pass from these equations to their linear approximations in order to obtain further results. Setting

$$\frac{dN_1}{dt} = \frac{dN_2}{dt} = 0$$

we obtain as the equilibrium population sizes

$$N_1 = e_2/a_{12} \equiv q_1, \qquad N_2 = e_1/a_{12} \equiv q_2.$$

We accordingly write

$$x_1 = N_1 - e_2/a_{12}, \qquad x_2 = N_2 - e_1/a_{12},$$

and neglecting products in $x_1 x_2$ as second order, we obtain as our linear approximation the equations

$$\frac{dx_1}{dt} = -e_2 x_2 = -a_{12}q_2 x_2,$$

$$\frac{dx_2}{dt} = e_1 x_1 = a_{12}q_1 x_1.$$

These equations yield

$$\frac{d^2 x_1}{dt^2} = -e_1 e_2 x_1 = -a_{12}^2 q_1 q_2 x_1;$$

the populations oscillate with a frequency $(e_1 e_2)^{1/2} = a_{12}(q_1 q_2)^{1/2}$: we have as solutions to these equations

$$x_1(t) = \frac{1}{\sqrt{e_1}} (A \cos \sqrt{e_1 e_2}\, t - B \sin \sqrt{e_1 e_2}\, t),$$

$$x_2(t) = \frac{1}{\sqrt{e_2}} (B \cos \sqrt{e_1 e_2}\, t + A \sin \sqrt{e_1 e_2}\, t),$$

where $A = \sqrt{e_1}\, x_1(0)$, $B = \sqrt{e_2}\, x_2(0)$.

1.1.3 From these solutions we can calculate the following time averages:

$$\lim_{T \to \infty} \frac{1}{2T} \int_{-T}^{T} x_1^2(t)\, dt = \frac{1}{2e_1}(A^2 + B^2) \equiv \sigma_1^2,$$

$$\lim_{T \to \infty} \frac{1}{2T} \int_{-T}^{T} x_2^2(t)\, dt = \frac{1}{2e_2}(A^2 + B^2) \equiv \sigma_2^2,$$

$$\lim_{T \to \infty} \frac{1}{2T} \int_{-T}^{T} x_1(t) x_2(t)\, dt = 0,$$

$$\lim_{T \to \infty} \frac{1}{2T} \int_{-T}^{T} x_1(t) \frac{dx_2(t)}{dt}\, dt = \lim_{T \to \infty} \frac{1}{2T} \int_{-T}^{T} \varepsilon_1 x_1^2(t)\, dt$$

$$= \tfrac{1}{2}(A^2 + B^2) = \sqrt{e_1 e_2}\, \sigma_1 \sigma_2 = a_{12} \sqrt{q_1 q_2}\, \sigma_1 \sigma_2.$$

Since we have

$$\lim_{T \to \infty} \frac{1}{2T} \int_{-T}^{T} x_1(t) \frac{dx_2(t)}{dt}\, dt$$

$$= \frac{d}{ds} \left[\lim_{T \to \infty} \frac{1}{2T} \int_{-T}^{T} x_1(t) x_2(t + s)\, dt \right]_{s=0}$$

this last relation permits us to evaluate the coefficient of interaction a_{12} from characteristics of the observed population curve.

1.1.4 The question may arise whether the covariance

$$\lim_{T \to \infty} \frac{1}{2T} \int_{-T}^{T} x_1(t) x_2(t+s) \, dt$$

is a sufficiently well-behaved function near $s = 0$ to permit finding its derivative. We may write

$$\lim_{T \to \infty} \frac{1}{2T} \int_{-T}^{T} x_1(t) x_2(t+s) \, dt$$

$$= \lim_{T \to \infty} \frac{1}{2T} \int_{-T}^{T} \frac{1}{a} (A \cos at - B \sin at)$$

$$\cdot [B \cos a(t+s) + A \sin a(t+s)] \, dt$$

where $a = \sqrt{e_1 e_2}$. Using the usual addition formulae, we write

$B \cos a(t+s) + A \sin a(t+s)$

$\quad = B(\cos at \cos as - \sin at \sin as)$

$\quad\quad + A(\sin at \cos as + \cos at \sin as).$

The value of our covariance is accordingly

$$\frac{1}{\sqrt{e_1 e_2}} [e_1 x_1^2(0) + e_2 x_2^2(0)] \sin \sqrt{e_1 e_2}\, s$$

which is no more difficult to handle than the observed population curve: indeed, much less so, since the integral should smooth out some of the noisy properties of an observed curve. Notice that for $s = 0$ the covariance is nil, while for $\sqrt{e_1 e_2}\, s$ about three the covariance is quite high. This phenomenon perhaps accounts for the time lag often observed in the response of a population's size to its food supply.

1.2.1 Complex predator-prey schemes: The Kerner formalism.

We now seek means of extending the results of the classical two-species case to communities of r species. The appropriate equations are now

$$\frac{dN_i}{dt} = e_i N_i + \sum_{i=1}^{r} a_{is} N_i N_s; \qquad a_{is} = -a_{si}.$$

This set of equations can be thought of as representing a food-web diagram: the species of the community are plotted as points on the diagram: those species with positive e_i, which presumably represent primary producers, are placed at the bottom of the diagram, and the diagram is drawn in such a way that energy tends to flow upward. A nonzero coefficient of interaction a_{is} is indicated by an

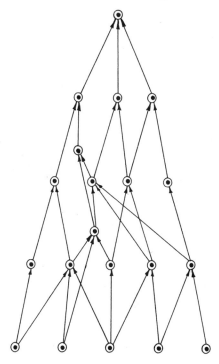

arrow connecting species i to species s, leading from i to s if a_{is} is negative: these arrows thus indicate the paths of energy flow in the community. In a more elaborate diagram the thicknesses of the arrows would be drawn proportional to the $a_{is}q_iq_s$, the rate of flow of mass between the species in question. All our analysis is an embroidery on this purely combinatorial frame.

1.2.2 We rewrite our community equations as

$$\frac{d \log N_i}{dt} = e_i + \sum_{s=1}^{r} a_{is}q_s(N_s/q_s)$$

where the q_i are equilibrium values of the N_i, values of the N_i for which all the $dN_i/dt = 0$. Setting $x_i = \log(N_i/q_i)$, we obtain

$$\begin{aligned}\frac{dx_i}{dt} &= e_i + \sum_{s=1}^{r} a_{is}q_s e^{x_s} \\ &= e_i + \sum_{s=1}^{r} a_{is}q_s + \sum_{s=1}^{r} a_{is}q_s(e^{x_s} - 1) \\ &= \sum_{s=1}^{r} a_{is}q_s(e^{x_s} - 1).\end{aligned}$$

In this form the equations are very easily integrated:

$$\sum_{i=1}^{r} q_i(e^{x_i} - 1)\frac{dx_i}{dt} = \sum_{i,s=1}^{r} a_{is}q_iq_s(e^{x_i} - 1)(e^{x_s} - 1) = 0,$$

$$\sum_{i=1}^{r} q_i(e^{x_i} - x_i) \equiv G = \text{const.}$$

Moreover, and this is what is of importance, our equations can be written in the form

$$\frac{dx_i}{dt} = \sum_s a_{is}\frac{\partial G}{\partial x_s}.$$

One visualizes a space consisting of all possible vectors $x_1, \ldots, x_r$: these equations of motion define a transformation of this space onto itself which transforms the vector $x_1, \ldots, x_r$ into the vector $U_t(x_1, \ldots, x_r)$ over the course of an interval of time t: one may think of these paths $U_t(x_1, \ldots, x_r)$ for the various possible vectors in the space, this set of trajectories, if you will, as representing lines of flow. This flow, this transformation, is volume-preserving: it sends any set into a set of equal volume: this is none other than Gibbs' theorem of conservation of extension in phase.

1.2.3 It is appropriate at this point to make some remarks concerning the handling of integrable systems of differential equations. Consider the system

$$\frac{dx_i}{dt} = \sum_s a_{is} \frac{\partial G}{\partial x_s}, \qquad a_{is} = -a_{si}.$$

If we consider our phase space as occupied by a fluid, the various portions of which follow the lines of flow defined by these equations, and if this fluid has a density $\rho(x_1, \ldots, x_r)$, we may apply the equation of continuity to this fluid:

$$\frac{d\rho}{dt} = -\sum_{i=1}^r \frac{\partial}{\partial x_i}\left(\rho \frac{dx_i}{dt}\right)$$

$$= \frac{\partial \rho}{\partial t} + \sum_{i=1}^r \frac{\partial \rho}{\partial x_i}\frac{dx_i}{dt} + \sum_{i=1}^r \rho \frac{\partial}{\partial x_i}\frac{dx_i}{dt} = 0.$$

This equation is no more than a book-keeper's statement, which says that the change in density of fluid in any little box is equal to the contribution from the material entering the box less the contribution from material originally in the box which then left. Using our original equations of motion, we have

$$\sum_{i=1}^r \rho \frac{\partial}{\partial x_i}\frac{dx_i}{dt} - \sum_{i,s} \rho a_{is} \frac{\partial^2 G}{\partial x_i \partial x_s} = 0.$$

Consequently, $d\rho/dt = 0$. Our flow is volume-preserving: if one adopts as the measure of a set of vectors in phase space the volume of this set, then this flow defines a measure-preserving transformation on our phase space. G remains unchanged under the flow defined by these equations: we have

$$\frac{dx_i}{dt} = \sum_s a_{is} \frac{\partial G}{\partial x_s}$$

and therefore

$$\sum_i \frac{\partial G}{\partial x_i} \frac{dx_i}{dt} = \frac{dG}{dt} = \sum_{i,s} a_{is} \frac{\partial G}{\partial x_i} \frac{\partial G}{\partial x_s},$$

thus our phase space of all possible vectors $(x_1, \ldots, x_r)$ decomposes into "phase shells" of constant G, each one left invariant under the measure-preserving flow U_t defined by our equations.

1.2.4 If we assume that none of these phase shells of constant G can be decomposed into two subshells of positive measure, both of which are invariant under the flow, then we can apply the ergodic theorem

$$\lim_{T \to \infty} \frac{1}{2T} \int_{-T}^{T} f(x_i(t))\, dt$$

$$= \frac{1}{S_C} \int_{G=c} f(x_i) \mu(d\sigma) \quad \text{almost everywhere}$$

where

$$\mu(dt) = d\sigma \Big/ \sqrt{\sum_i (\partial G/\partial r_i)^2}$$

is the "microcanonical" measure induced on the surface element $d\sigma$ of the phase shell by the volume measure in phase space, and S_C is the area of the phase shell $G = c$: this theorem permits us to evaluate the time average of a quantity over almost any given trajectory by an average over phase space.

If we consider a small subset $x_1, \ldots, x_k$, k much less than r, of our phase variables, and if G can be written as $\sum_i G_i(x_i)$, then it can be shown that the measure $\mu(d^k x)$ of an element in this hyperplane of phase space is equal to

$$\mu(d^k x) = c \exp\left[-\frac{1}{\theta} \sum_{i=1}^{k} G_i(x_i)\right]$$

where c is a normalising factor and θ is some constant dependent upon the value of G. The ergodic theorem is also appropriate to averages of functions of these variables taken with this measure, and accordingly this measure, the so-called 'canonical average', occupies a central place in all applications of this statistical technique.

1.2.5 Although topological considerations will play little explicit role in this paper, the topological nature of our phase shells of constant G is of the utmost importance to us. If our phase shell is not a connected surface, we have no reason to expect the ergodic theorem to apply to it as a whole, although it may reasonably be expected to apply to each of the pieces separately. Some example of the effect of purely topological considerations can be seen from the following remark: if we are dealing with a system of two equations, then, if the figure $G(x_1, x_2) = c$ is topologically equivalent to a circle, a population with this value of G will execute a uniquely defined periodic cycle: if it is topologically equivalent to several nonintersecting circles, the population will exhibit one of several possible forms of periodic behavior, the particular form depending on initial conditions. On the other hand, if $G(x_1, x_2) = c$ is not a closed figure, the solutions to the equations will not be periodic: if $G = x^2 - y^2$, our equations and solutions will be

$$dx/dt = -2ay, \qquad x = x_0 \cosh 2at - y_0 \sinh 2at,$$
$$dy/dt = -2ax, \qquad y = y_0 \cosh 2at - x_0 \sinh 2at.$$

If we move to higher dimensional G-surfaces, the geometry becomes less restrictive: there is nothing in the nature of a multidimensional closed G-surface which forces the trajectory to return to its initial point, and thus, for closed G-surfaces, the two-equation theorem of population fluctuations is replaced by the multidimensional.

POINCARE RECURRENCE THEOREM. *A trajectory will almost always return infinitely often to any neighborhood of positive measure of its initial point.*

PROOF. Consider a set g_0 of positive measure in the shell $G = c$, let g_t be the set $U_t(g_0)$ into which g_0 is transformed in time t, and let Γ_0 be the union over all finite t greater than or equal to zero of the g_t: Γ_0 will be of finite measure if the shell is a closed surface (topology again!). Similarly, let Γ_{t_1} be the union over all finite t greater than or equal to some positive t_1 of the g_t: obviously the set Γ_{t_1} is contained in the set Γ_0. But $\Gamma_{t_1} = U_{t_1}(\Gamma_0)$, so that, by the theorem of conservation of extension in phase, the measures of the two sets are equal, so that Γ_{t_1} must contain all of Γ_0 and consequently all of g_0, save possibly a set of measure zero.

It is an easy step to see that the trajectory must return not just once, but infinitely often, to this neighborhood g_0, so that if we place our initial point in this neighborhood, our theorem is proved.

1.2.6 We now give some examples of the working of the method. Consider the set of equations

$$\frac{dx_i}{dt} = \sum_{s=1}^{v} a_{is} x_s :$$

if we write $G = \sum_{i=1}^{v} \frac{1}{2}x_i^2$, then these equations assume the form
$$\frac{dx_i}{dt} = \sum_{s=1}^{v} a_{is} \frac{\partial G}{\partial x_s}.$$

First we evaluate some phase averages: by some unattractive juggling involving the divergence theorem we find that
$$\int_{G=c} x_i \frac{\partial G}{\partial x_i} d\sigma \Big/ \sqrt{\sum_s \left(\frac{\partial G}{\partial x_s}\right)^2} = V_c$$
where V_c is the volume enclosed by the shell $G = c$: remembering that
$$\frac{\partial G}{\partial x_i} = x_i$$
we find, using the ergodic theorem, that
$$\lim_{T \to \infty} \frac{1}{2T} \int_{-T}^{T} x_i^2(t) \, dt = V_c/S_c,$$
where S_c is the area of the surface $G = c$: we find, in particular, that the time average of all the x_i^2's are the same. It can likewise be shown that
$$\int_{G=c} \frac{\partial G}{\partial x_i} d\sigma \Big/ \sqrt{\sum_s \left(\frac{\partial G}{\partial x_s}\right)^2} = 0,$$
the time averages of the x_i are all zero.

If we consider the variable x_1 in the case where we have a large number of equations, we may invoke the canonical average theorem of 1.2.4 to conclude that it follows a distribution law of $(1/\sqrt{2\pi\theta}) \, e^{-x_i^2/2\theta}$ which is none other than the Gaussian distribution with variance θ: our considerations above indicate that $\theta = V_c/S_c$. It can be shown by other methods that the solution to these equations is a superposition of sine-waves: this is therefore the shortest proof I know of the fact that the sum of a large number of sine waves follows a Gaussian distribution.

One other average of interest is that of

$$\int x_i \frac{dx_s}{dt} e^{-G/\theta} d^n x = \int x_i \sum_r a_{sr} \frac{\partial G}{\partial x_r} e^{-G/\theta} d^n x$$
$$= \int a_{si} x_i^2 e^{-x_i^2/2\theta} dx_i = a_{si}\theta.$$

Since we have

$$\frac{d}{ds}\left[\lim_{T\to\infty} \frac{1}{2T} \int_{-T}^{T} x_i(t) x_s(t+s)\, dt\right]_{s=0} = \int x_i \frac{dx_s}{dt} e^{-G/\theta} d^n x,$$

this gives us a relation which enables the measuring of the a_{is} from output curves.

1.2.7 Our second example shall be a thumbnail sketch of classical statistical mechanics. We consider a set of particles in a box, whose energy of interaction and whose potential energy due to gravitation are much less than their kinetic energy. The equations of motion of our system are the canonical equations of Hamilton:

$$dp_i/dt = -\partial H/\partial q_i, \qquad dq_i/dt = \partial H/\partial p_i,$$

where q_i is a co-ordinate and p_i its conjugate momentum: $H = \sum_i p_i^2/2m$ is the energy of our system. Here our matrix a_{is} is that type matrix

$$\begin{pmatrix} 0 & -1 & 0 & 0 & & & & \\ 1 & 0 & 0 & 0 & & & & \\ 0 & 0 & 0 & -1 & & \mathbf{O} & & \\ 0 & 0 & 1 & 0 & & & & \\ & & & & & & 0 & -1 \\ & & \mathbf{O} & & & & 1 & 0 \end{pmatrix}$$

which represents the normal form of all skew-symmetric matrices.

Since the particles are almost always freely moving, dp_i/dt is almost always nearly zero and dq_i/dt is equal to p_i/m, where m is the mass of the particle, when we employ rectangular co-ordinates. In these co-ordinates, we find in the same manner as before that the averages of

$$p_i \frac{\partial H}{\partial p_i} = \frac{p_i^2}{m}$$

are all the same: the particles of our system all have the same average energy. This is the famous equipartition theorem of statistical mechanics.

Using the theorem of micro-canonical averages, we find that the probability distribution of p_i is

$$ce^{-H(p_i)/\theta} = ce^{-p_i^2/2m\theta}.$$

Here, θ is proportional to temperature: the constant of proportionality is Boltzmann's constant: thus we arrive at the Maxwell-Boltzmann distribution for velocities. Similarly, the positions of the particles are uniformly distributed through the box, as we would expect.

1.3.1 Applications of the Kerner formalism: Full theory.
Having developed this apparatus, we shall now proceed to apply it to complex predator-prey schemes. Our equations are

$$\frac{dx_i}{dt} = \sum_s a_{is} \frac{\partial G}{\partial x_s}$$

where $G = \sum_i q_i(e^{x_i} - x_i)$: $x_i = \log N_i/q_i$, where N_i is the population of species i and q_i is the equilibrium population of species i, written in units of mass.

Evaluating the averages

$$\frac{1}{S_c}\int_{G=c} \frac{\partial G}{\partial x_i}\,d\sigma \Big/ \sqrt{\sum_r \left(\frac{\partial G}{\partial x_r}\right)^2}$$

and

$$\frac{1}{S_c}\int x_i \frac{\partial G}{\partial x_i}\,d\sigma \Big/ \sqrt{\sum_r \left(\frac{\partial G}{\partial x_r}\right)^2}$$

as in 1.2.6, we find from the first one, that the average of $q_i(N_i/q_i - 1)$ is 0, so that, using the ergodic theorem, we conclude that the time average of N_i is q_i: from the second average, again with the customary application of the ergodic theorem, we conclude that the time averages of

$$x_i \frac{\partial G}{\partial x_i} = q_i\left(\log \frac{N_i}{q_i}\right)\left(\frac{N_i}{q_i} - 1\right)$$

are the same for all species. The significance of this result appears when we evaluate the average of $x_i\,\partial G/\partial x_i$ using the canonical average: we find through an integration by parts, that

$$c\int x_i \frac{\partial G}{\partial x_i}\,e^{-G/\theta}\,d^k x = \theta$$

(c is the customary normalizing constant, and *not* a value of G!). We have thus established an equipartition theorem for this parameter which appears in the canonical distribution.

1.3.2 We turn now to the canonical distribution, which we use both as a measure for subspaces of phase space of relatively small dimension, and as an approximate measure for the microcanonical measure, which is difficult to handle. As in the microcanonical case,

$$c\int \frac{\partial G}{\partial x_i}\,e^{-G/\theta}\,d^k x = 0.$$

We also find that

$$c\int\left(\frac{\partial G}{\partial x_i}\right)^2 e^{-G/\theta}\, d^k x = c'\int_{-\infty}^{\infty}\left(\frac{\partial G_i}{\partial x_i}\right)^2 e^{-G_i/\theta}\, dx$$

$$= c'\left\{-\theta\left[\frac{\partial G}{\partial x_i} e^{-G_i/\theta}\right]_{-\infty}^{\infty} + \theta\int_{-\infty}^{\infty} e^{-G_i/\theta}\frac{\partial^2 G}{\partial x_i^2}\, dx_i\right\}$$

where $G_i = q_i(e^{x_i} - x_i)$ and c' is a normalizing constant for $e^{-G_i/\theta}$. Since $\partial^2 G/\partial x_i^2 = \partial G/\partial x_i + q_i$, we obtain, with a little more effort, that this integral is equal to θq_i. Substituting for $\partial G_i/\partial x_i$ in terms of N_i and q_i, this result tells us that the time average of

$$\frac{(N_i - q_i)^2}{q_i}$$

is the same for all species in the community, so that if we plot the variance of population size against population size for each species of the community, we should get a straight line through the origin with slope equal to θ when the theory applies. This law is strangely reminiscent of the $\sqrt{N}$ laws one encounters in many applications of probabilistic reasoning. As we are dealing here with masses rather than numbers, the significance is somewhat more obscure.

1.3.3 According to the theorem of the canonical average, x_i is distributed according to the law

$$C \exp\left[-\frac{q_i}{\theta}(e^{x_i} - x_i)\right].$$

Thus N_i is distributed according to the law

$$P(N_i) = \frac{1}{\Gamma(q_i/\theta)}\, \theta^{-q_i/\theta} N_i^{q_i/\theta - 1} e^{-N_i/\theta}.$$

Kerner [2] has verified this distribution for Arctic foxes on the basis of Canadian fox catch records, but as we shall see later, the verification of the distribution does not necessarily verify the scheme of interactions postulated here.

Kerner has also remarked that if we have an array of species of the same average population size, and if one samples the community randomly in such a fashion that the probability of collecting n specimens of species r is $e^{-tN_r}[(tN_r)^n/n!]$, then the distribution of catch sizes among the various species of the community will be given by the law

$$f(n) = \int_0^\infty P(N) e^{-tN} \frac{(tN)^n}{n!} dN$$

$$= \frac{\Gamma(n + q_r/\theta)}{n!\,\Gamma(q_r/\theta)} \left[\frac{\theta t}{1 + \theta t}\right]^{n+q_r/\theta} (\theta t)^{-q_r/\theta}.$$

If we take the limit $q_r/\theta \to 0$, we obtain the distribution of catch sizes of tropical moths discussed by Fisher, Corbett and Williams [3].

Taking a large value of θ for tropical communities, however, seems entirely unreasonable: Biologists always speak of the *stability* of the tropics. The large variance observed must be attributed to striking differences in the population means.

1.3.4 We next evaluate the average of $x_i\, dx_r/dt$.

$$c \int x_i \frac{dx_r}{dt} e^{-G/\theta} d^k x = c \sum_s a_{rs} \int x_i \frac{\partial G}{\partial x_s} e^{-G/\theta} d^k x$$

$$= c \sum_s a_{rs} \int x_i \frac{\partial G_s}{\partial x_s} e^{-G/\theta} d^k x$$

$$= c' a_{ri} \int x_i \frac{\partial G_i}{\partial x_i} e^{-G_i/\theta} dx_i = a_{ri} \theta$$

since x_i is distributed independently of any function of x_s. We thus have

$$a_{ri}\theta = c\int x_i \frac{dx_r}{dt} e^{-G/\theta} d^k x$$

$$= \lim_{T\to\infty} \frac{1}{2T} \int_{-T}^{T} x_i \frac{dx_r}{dt} dt$$

$$= \frac{d}{ds} \left[\lim_{T\to\infty} \frac{1}{2T} \int_{-T}^{T} x_i(t) x_r(t+s) dt \right]_{s=0}.$$

This result is the one we enunciated in linear approximation for the two-species case in 1.1.3. Whereas in the two-species case it was much easier to determine the a_{is} from the frequency of the periodic fluctuations, in many-species systems, correlation methods provide the only reasonable means of measuring these coefficients. The existence of nonzero derivatives for these correlations seems remarkable in view of the fact that $x_i(t)$ and $x_s(t)$ are not only uncorrelated, but independently distributed of each other.

1.4.1 Applications of the Kerner formalism: Linear approximations. We return to the Volterra equations

$$\frac{dN_i}{dt} = N_i \left(e_i + \sum_s a_{is} N_s \right).$$

Substituting in $N_i = q_i + v_i$, we obtain

$$\frac{dv_i}{dt} = (q_i + v_i)\left(e_i + \sum_s a_{is} q_s + \sum_s a_{is} v_s \right).$$

We thereby obtain

$$\frac{dv_i}{dt} = (q_i + v_i)\left(\sum_s a_{is} v_s \right),$$

whose linear approximation is obviously

$$\frac{dv_i}{dt} = \sum_s q_i a_{is} v_s.$$

If we set $G = \sum_i v_i^2/2q_i\theta$, we may write

$$\frac{dv_i}{dt} = \sum_s a_{is} q_i q_s \frac{\partial G}{\partial v_s};$$

the coefficients are the rates of mass flow between the species in question. The probability distribution $P(v_i)$ of v_i is given by the theorem of the canonical average, and is

$$P(v_i) = \frac{1}{\sqrt{2\pi q_i\theta}} \exp\left[-v_i^2/2q_i\theta\right];$$

the variance in population size is found here, as in the full theory, to be $q_i\theta$.

1.4.2 We now seek to evaluate the frequency with which the population curve $v_i(t)$ crosses the line $v_i = a$, or, in other terms, the frequency of zeroes of the function $v_i(t) - a$. Consider an arbitrary function $F(t)$, and consider the Dirac delta-function $\delta(t)$ whose property it is to be zero for $t \neq 0$, and yet to be so strongly infinite at $t = 0$ that

$$\int_{-\infty}^{\infty} \delta(t)\, dt = 1,$$

then for any small ϵ-interval

$$\int_{-\epsilon}^{\epsilon} \delta(F(t))\, |F'(t)|\, dt$$

will be 1 if the interval contains a zero of $F(t)$ and zero otherwise, and the frequency of zeroes per unit time will be given by

$$\lim_{T \to \infty} \frac{1}{2T} \int_{-T}^{T} \delta(F(t))\, |F'(t)|\, dt.$$

Thus we require to evaluate

$$\lim_{T\to\infty} \frac{1}{2T} \int_{-T}^{T} \delta[v_i(t) - a] \left|\frac{dv_i}{dt}\right| dt = c \int \delta[v_i(t) - a] \left|\frac{dv_i}{dt}\right|$$

$$\exp\left[\sum_r \frac{v_r^2}{2q_r\theta}\right] d^k v$$

(c is our customary normalizing constant). This last integral is equal to

$$\frac{1}{\sqrt{2\pi q_i \theta}} \exp\left[\frac{-a^2}{2q_i\theta}\right] c' \int \left|\frac{dv_i}{dt}\right| \exp\left[-\sum_r' \frac{v_r^2}{2q_r\theta}\right] d^{k-1}v,$$

where the primed summation excludes the v_i term: since $|dv_i/dt|$ does not depend on v_i, we may write this as

$$\frac{1}{\sqrt{2q_i\pi\theta}} \exp\left[\frac{-a^2}{2q_i\theta}\right] c \int \left|\frac{dv_i}{dt}\right| \exp\left[-\sum_r \frac{v_r^2}{2q_r\theta}\right] d^{k-1}v.$$

1.4.3 This section shall be concerned with the evaluation of the integral

$$c \int \left|\frac{dv_i}{dt}\right| \exp\left[-\sum_r \frac{v_r^2}{2q_r\theta}\right] d^k v.$$

First we write

$$\left|\frac{dv_i}{dt}\right| = \frac{1}{\pi} \int_{-\infty}^{\infty} \frac{ds}{s^2} \left[1 - \cos\frac{dv_i}{dt}s\right].$$

By virtue of this relation our integral becomes

$$c \int d^k v \exp\left[-\sum_r \frac{v_r^2}{2q_r\theta}\right] \frac{1}{\pi} \int \frac{ds}{s^2}\left[1 - \cos\frac{dv_i}{dt}s\right]$$

$$= \frac{c}{\pi} \int \frac{ds}{s^2} d^k v \exp\left[-\sum_r \frac{v_r^2}{2q_r\theta}\right]\left(1 - \cos\frac{dv_i}{dt}s\right).$$

We have

$$\int \exp\left[-\sum_r \frac{v_i^2}{2q_r\theta}\right] d^k v = \frac{1}{c}$$

and

$$\int \exp\left[-\sum_r \frac{v_r^2}{2q_r\theta}\right] \cos\left(\frac{dv_i}{dt}s\right) d^k v$$

$$= \int \exp\left[-\sum_r \frac{v_r^2}{2q_r\theta} + is\frac{dv_i}{dt}\right] d^k v$$

(the normal distribution is even!). Substituting $dv_i/dt = \sum_s q_i a_{is} v_s$, we obtain

$$\int \exp\left[-\sum_r \left[\frac{v_r^2}{2q_r\theta} + isq_i a_{ir} v_r\right]\right] d^k v.$$

We may write

$$\frac{v_r^2 + 2isq_i q_r a_{ir}\theta v_r}{2q_r\theta} = \frac{(v_r + isq_i q_r a_{ir}\theta)^2}{2q_r\theta} + \tfrac{1}{2} q_i^2 q_r a_{ir}^2 \theta.$$

Using this relation, we obtain for

$$\int \exp\left[-\sum_r \left[\frac{v_r^2}{2q_r\theta} + isq_i a_{ir} v_r\right]\right] d^k v$$

the value

$$\frac{1}{c} \exp\left[-\sum_r q_i^2 q_r a_{ir}^2 \theta\right].$$

Accordingly, the integral we seek becomes

$$\frac{1}{c} \int_{-\infty}^{\infty} \frac{ds}{s^2} \left(1 - \exp\left[-\sum_r q_i^2 q_r a_{ir}^2 \theta\right]\right)$$

whose value is

$$\sqrt{\frac{2}{\pi c^2}} \sqrt{\sum_r q_i^2 q_r a_{ir}^2 \theta}.$$

1.4.4 Using the results at the ends of 1.4.2 and 1.4.3, we find that the frequency of zeroes of the function $v_i(t) - a$ is

$$\frac{1}{\pi} \exp\left[-\frac{a^2}{2q_i\theta}\right]\sqrt{\sum_r q_i q_r a_{ir}^2}.$$

One might argue that the appropriate measure of population fluctuations in a species is in terms of their ratio to the equilibrium population size: if we accordingly substitute $v_i/q_i = x_i$: the frequency of zeroes of the curve $x_i(t) - a$ is

$$\frac{1}{\pi} \exp\left[-\frac{q_i a^2}{2\theta}\right]\sqrt{\sum_r q_i q_r a_{ir}^2}.$$

If we minimize the average of

$$\sum_{r=1}^{k} a_{ir}^2 q_i q_r, \quad \text{which is} \quad \frac{1}{k}\sum_{i,r=1}^{k} a_{ir}^2 q_i q_r$$

(k is the number of species in the community: we have irrevocably crossed over to the point of view that the canonical average is an approximation to the microcanonical average), we will minimize, at least to a rough approximation, the frequency of gross fluctuations in population size, so that the values of a_{is} which minimize this average will specify, in some sense, the community structure of optimal stability. We find it appropriate to minimize $\sum a_{is}^2 q_i q_s$ under the restriction that the productivity of the community is equal to P. We define the productivity of a community to be the total rate of mass flow through its various trophic levels: since $|a_{is}| q_i q_s$ is the rate of mass flow between species i and species s, the productivity P is given by

$$\frac{1}{2}\sum |a_{is}| q_i q_s$$

where the sum is extended over all the possible species pairs in the community.

We perform the required minimization by the method of Lagrange multipliers: we accordingly seek to minimize

$$\sum_{i,s} |a_{is}|^2 q_i q_s - \lambda \left(\sum_{i,s} |a_{is}| q_i q_s - 2P \right)$$

with respect to λ and the $|a_{is}|$. The first condition yields

$$\frac{\partial}{\partial \lambda} \left[\sum_{i,s} |a_{is}|^2 q_i q_s - \lambda \left(\sum_{i,s} |a_{is}| q_i q_s - 2P \right) \right] = 0$$

or

$$\sum_{i,s} |a_{is}| q_i q_s = 2P,$$

and the second yields

$$2 |a_{is}| q_i q_s = \frac{\lambda}{2} q_i q_s = 0, \qquad |a_{is}| = \frac{\lambda}{4}.$$

Since

$$\tfrac{1}{2} \sum |a_{is}| q_i q_s = P = \frac{\lambda}{8} \left[\left(\sum_i q_i \right)^2 - \sum q_i^2 \right] \quad (a_{ii} = 0!),$$

and since, if the community is reasonably diverse, we may set

$$(\sum q_i)^2 - \sum q_i^2 \approx (\sum q_i)^2 = B^2;$$

where B is the total biomass of the community, we may write

$$\lambda = \frac{8P}{B^2}, \qquad |a_{is}| = \frac{2P}{B^2}$$

and

$$\frac{1}{k} \sum |a_{is}|^2 q_i q_s = \frac{1}{k} \frac{4P^2}{B^2}.$$

The average frequency of population crashes per species per unit time is therefore proportional, for this type of community structure, to

$$\frac{2}{\sqrt{k}} \frac{P}{B} e^{-q_i a^2/2\theta}$$

Thus community stability increases with a proportional increase of productivity and biomass, or with increase in the connectedness of the food web: it decreases if productivity is increased for a fixed biomass.

1.4.5 This result was derived in a most abstruse and imprecise way, but it admits of a simple interpretation. We may think of P/B as a "turnover rate" of the community, the rate at which living matter is renewed. The higher this turnover rate, the less stable the community, for populations are more likely then to "overshoot" after disturbance. The term reflects the fact that, other things being equal, a population is the stabler the larger its average value. As we shall see later, θ is a measure of the physical environment: the higher θ, the less stable the environment. Finally, a species population is the stabler the more species it feeds on, for it is the less affected by fluctuations in any one of them. Notice that productivity and biomass are defined in terms of herbivore equivalents: this definition emphasizes the contribution of higher trophic levels far more than does the usual definition in terms of energy or weight.

1.4.6 How do we attach meaning to the stability theorem? What do we expect it to tell us?

To answer this, we ask how a community is organised. Its principle of organisation follows from its manner of development, just as the principle that natural selection organises an animal to perpetuate its kind follows from the nature of inheritance in animals. A community develops through a series of successive invasions, successive colonisations: we see the process take place when a field is abandoned. Weeds spring up, then shrubbier things, all from seeds which have drifted in one way or another. Finally, after many years, a forest of stable composition develops, stable precisely because it is the association most resistant to invasion from the outside.

It follows that each species specialises to as narrow a way of life as is consistent with survival, for if not it will be replaced by species more specialised, and therefore more effective. Species populations are thus likely to be equally stable everywhere: the theorem, then, sets limits on the possible degree of specialisation. In an unstable environment, creating a high turnover rate, species have to be more diversified in habit than under less disturbed conditions. One may expect remarkably low turnover rates in stable but unproductive settings. Other observations also follow. In a stable environment, populations will likely be smaller: so, therefore, will be their fluctuations. Spectacular insect outbreaks, so familiar in forests of the north, are not a part of tropical forest biology [4].

References: Part I

1. I have done nothing in this chapter but apply the calculations of Volterra and Kerner to various biological problems, a few of which they had not envisioned.

2. E. H. Kerner, *Further considerations on the statistical mechanics of biological associations*, Bull. Math. Biophysics **21** (1959), 217–255.

3. R. A. Fisher, H. S. Corbett and C. B. Williams, *The relation between the number of species and the number of individuals in a random sample of an animal population*, J. Animal Ecology **12** (1943), 42–58.

4. C. Elton, *The ecology of invasions by animals and plants*, Methuen, London, 1958.

II. THE TOTAL ENVIRONMENT AS A SOURCE OF RANDOM NOISE

2.1 The Maxwell-Boltzmann distribution can be derived two ways. We may assume that we are dealing with an unperturbed, conservative mechanical system and use the formalism of statistical mechanics to show that the distribution through time of the velocity v_i of particle i is Gaussian. Or we may assume that the particle moves in a viscous medium, so that its velocity tends to decline geometrically, but that it is subject to random shocks corresponding to a Gaussian Markov input. Is there a broader analogy between conservative systems and Gaussian Markov processes?

2.2.1 We have shown that if r is sufficiently large, the system of equations

$$\frac{dx_i}{dt} = \sum_{s=1}^{r} a_{is} \frac{\partial G}{\partial x_s} \quad \left[G = \sum_{i=1}^{r} G_i(x_i) \right]$$

has the property that the distribution through time of $x_i(t)$ is of the form $c \exp(-G_i/\theta)$, where θ is a positive constant.

Let us now consider the stochastic equation

$$dx_i(t, a) = -b \, \partial G_i/\partial x_i + dB(t, a),$$

$dB(t, a)$ is a term whose integral over the time interval $t_1 < t < t_2$ is a random variable with mean 0, variance $\sigma^2(t_2 - t_1)$, and a Gaussian distribution. a is an index of the possible paths $x(t)$, varying between 0 and 1, such that the probability

$$A < \int_{t_1}^{t_2} dB(t, a) \leq B$$

is equal to the measure of the set of a for which this

equation holds. We further assume that the random variables so obtained for nonoverlapping time intervals are independent. Mathematicians call $dB(t, a)$ a Gaussian Markov input, physicists a Langevin force term.

2.2.2 There are a variety of ways to solve such a stochastic equation. Perhaps the easiest is to derive an equation in the probability density $\varphi(x, t)$ of x as a function of the time t. If future change in x depends only on the present value and not also on the past, we may write

$$\varphi(x, t + \Delta t) = \int \varphi(x - a, t) f(x - a, x, \Delta t) \, da$$

where $f(x, x - a, \Delta t)$ is the probability of a shift from $x - a$ to x in time Δt. We may expand

$$\varphi(x - a, t) f(x - a, x, \Delta t)$$

as a Taylor series about x, obtaining

$$\varphi(x, t) f(x, x + a, \Delta t) - a \frac{\partial}{\partial x} [\varphi(x, t) f(x, x + a, \Delta t)]$$
$$+ \frac{a^2}{2} \frac{\partial^2}{\partial x^2} [\varphi F].$$

Substituting the Taylor series under the integral, we obtain

$$\int \varphi(x - a, t) f(x - a, x, \Delta t) \, da$$
$$= \varphi(x, t) \int f(x, x + a, \Delta t) \, da$$
$$- \frac{\partial}{\partial x} \left[\varphi(x, t) \int a f(x, x + a, \Delta t) \, da \right]$$
$$+ \frac{1}{2} \frac{\partial^2}{\partial x^2} \left[\varphi(x, t) \int a^2 f(x, x + a, \Delta t) \, da \right].$$

Since x must shift to some value during the time interval Δt,
$$\int f(x, x+a, \Delta t)\, da = 1.$$
$$\int a f(x, x+a, \Delta t) = -b \frac{\partial G_i}{\partial x_i} \Delta t,$$
where $-b\, \partial G_i/\partial x_i$ is the expected rate of change of x_i.
$$\int a^2 f(x_i, x_i+a, \Delta t)\, da = \sigma^2\, \Delta t,$$
the mean square change in x_i during time Δt. From whence we obtain
$$\varphi(x_i, t+\Delta t) - \varphi(x_i, t) = \Delta t \left[\frac{\sigma^2}{2} \frac{\partial^2 \varphi}{\partial x_i^2} - \frac{\partial}{\partial x_i}\left(-b \frac{\partial G_i}{\partial x_i} \varphi\right) \right],$$
$$\frac{\partial \varphi}{\partial t} = \frac{\sigma^2}{2} \frac{\partial^2 \varphi}{\partial x_i^2} + \frac{\partial}{\partial x_i}\left[b \frac{\partial G_i}{\partial x_i} \varphi \right].$$

As time wears on, an equilibrium distribution is approached, for which $\partial \varphi/\partial t = 0$. The equilibrium distribution satisfies the equation
$$\frac{\sigma^2}{2} \frac{\partial \varphi}{\partial x_i} = -b\left(\frac{\partial G_i}{\partial x_i} \varphi \right)$$
$$\frac{1}{\varphi} \frac{\partial \varphi}{\partial x_i} = -\frac{2b}{\sigma^2} \frac{\partial G_i}{\partial x_i}$$
from whence we make bold to conclude
$$\varphi(x_i) = c \exp\left[(-2b/\sigma^2) G_i(x_i) \right].$$

2.3 We thus find that x_i has the same distribution whether it is a component of a conservative system satisfying the equations
$$\frac{dx_i}{dt} = \sum_{s=1}^{r} a_{is} \frac{\partial G_s}{\partial x_s}$$
or whether it satisfies a stochastic equation

$$dx_i(t, a) = -b \frac{\partial G_i}{\partial x_i} dt + dB(t, a).$$

The effects of the other components on x_i in a conservative system are to this extent analogous to the effects of a Gaussian Markov input.

2.4 In particular, we have shown that if the dynamical equations of a community are

$$\frac{d \log N_i}{dt} = e_i + \sum_{s=1}^{r} a_{is} N_s$$

then N_i will follow the probability density

$$cN_i^{q_i/\theta - 1} e^{-N_i/\theta}.$$

In particular, we showed this by setting $x_i = \log N_i/q_i$, where q_i is the equilibrium value of N_i, writing

$$\frac{dx_i}{dt} = e_i + \sum_s a_{is} q_s e^{x_s} = e_i + \sum_s a_{is} q_s + \sum_s a_{is} q_s (e^{x_s} - 1)$$

and showing that

$$\frac{dx_i}{dt} = \sum a_{is} \frac{\partial G_s}{\partial x_s}, \quad \text{where} \quad G = \sum q_i (e^{x_i} - x_i).$$

We could, however, consider a logistically regulated population in a random environment. Then we could express the expected rate of change of $\log N_i$ as

$$\frac{d \log N_i}{dt} = e_i - aN_i = a(q_i - N_i).$$

If we set $x_i = \log N_i/q_i$, we have

$$\frac{dx_i}{dt} = aq_i(1 - e^{x_i}) = -a \frac{\partial G}{\partial x_i}$$

where $G_i = q_i(e^{x_i} - x_i)$. If we suppose that the rate of multiplication of the population is subject to random environmental shocks taking the form of a Gaussian Markov input so that

$$dx_i(t, a) = -a\frac{\partial G_i}{\partial x_i} dt + dB(t, a),$$

we find x_i has the distribution

$$c \exp\left[-\frac{2a}{\sigma^2}(e^{x_i} - x_i)\right]$$

so that N_i has the distribution

$$cN_i^{2q_ia/\sigma^2 - 1} e^{-2N_i a/\sigma^2}.$$

The probability distribution of N_i thus tells us little about the mechanism of population regulation. The population of Arctic foxes [2] in Canada follows this distribution, judging by the records of fox catches in that country, but it might be regulated either by crowding or by food web relationships.

2.5.1 How do we decide whether a population of Arctic foxes is regulated by crowding or by food web relationships? More generally, how can one use observed population curves to determine the regulatory mechanisms of these populations? The Volterra theory suggests that the answer is to be found in the derivatives of correlations. In particular, we showed that

$$\frac{d}{ds}\left[\lim_{T\to\infty} \frac{1}{2T}\int_{-T}^{T} x_i(t)_r(t+s)\, dt\right]_{s=0} = a_{ir} q_r \theta\left(\frac{1}{q_r}\right)$$

where q_s is the average population size of species s and $q_s\theta$ is its variance. How does the correlation behave in the logistic case?

2.5.2 Let us treat the linear approximation of the "stochastic logistic." This will be

$$dx(t, a) = -aKx(t, a)\, dt + dB(t, a).$$

What we mean by such a "stochastic differential equation" is that

$$\int_0^s f(t)\,dx(t,a) = -aK\int_0^s x(t,a)f(t)\,dt + \int_0^s f(t)\,dB(t,a)$$

for almost every function $f(t)$. We substitute for

$$\int_0^s f(t)\,dx(t,a)$$

the expression

$$f(s)x(s,a) - f(0)x(0,a) - \int_0^s f'(t)x(t,a)\,dt$$

and obtain

$$x(s,a) = \frac{1}{f(s)}\left\{\int_0^s [f'(t) - aKf(t)]x(t,a)\,dt + f(0)x(0,a) + \int_0^s f(t)\,dB(t,a)\right\}.$$

If we set $f(t) = e^{aKt}$, so that $f'(t) = aKf(t)$, then we find

$$x(s,a) = e^{-aKs}x(0) + e^{-aKs}\int_0^s e^{aKt}\,dB(t,a).$$

A function $\int_0^s f(t)\,dB(t,a)$ has a Gaussian distribution with mean 0 and variance $\int_0^s f^2(t)\,dt$. Therefore $x(s,a)$ has a Gaussian distribution with mean $e^{-aKs}u(0)$ and variance

$$\frac{\sigma^2}{2Ka}(1 - e^{-2aKs}).$$

2.5.3 For s greater than zero,

$$\lim_{T\to\infty}\frac{1}{T}\int_{-T}^T x(t,a)x(t+s,a)\,dt$$

$$= \int_0^1 x(t,a)e^{-aKs}x(t,a)\,da = \frac{\sigma^2}{2aK}e^{-aKs}.$$

For s less than zero the result is

$$\frac{\sigma^2}{2aK}e^{-aK|s|}.$$

Thus, for a logistically regulated species in a random environment, the autocorrelation

$$r(s) = \frac{1}{2T} \int_{-T}^{T} x(t, a) x(t + s, a) \, dt$$

exhibits a sharp peak as $s = 0$ and falls off exponentially at each side of the peak. For a species living in a stable environment and limited purely by food web relationships, the autocorrelation function $r(s)$ is smooth at $s = 0$, and resembles the maximum of a parabolic curve.

Notice, however, that

$$\left. \frac{d}{ds} r(s) \right|_{0+} = \lim_{\substack{h > 0 \\ h \to 0}} \left[\frac{r(h) - r(0)}{h} \right] = -\frac{\sigma^2}{2} = -\frac{\sigma^2}{2aK} (aK).$$

In analogy with our result from the Volterra theory, the derivative of this autocorrelation is equal to the population mean K, times the variance $\sigma^2/2aK$, times a coefficient a indicating the effect of population size on population growth rate.

REFERENCES: PART II

1. I have made great use of the theory of Gaussian Markov processes, using both stochastic equations and diffusion methods. The clearest account of stochastic equations is in

J. L. Doob, *The Brownian movement and stochastic equations*, Ann. of Math. **43** (1942), 351–369.

I have also made use of techniques described in

N. Wiener, *Nonlinear problems in random theory*, Wiley, New York, and M.I.T. Press, Cambridge, Mass., 1958.

2. E. H. Kerner, *Further considerations on the statistical mechanics of biological associations* Bull. Math. Biophysics **21** (1959), 217–255.

III. Correlations and Their Uses

3.1.1 In general, food web relationships, crowding effect and environmental noise all play a role in population regulation. Can we incorporate all these into a general theory?

3.1.2 Let us begin with a study of the dynamical equations

$$dx_i(t, a) = \sum_{s=1}^{r} a_{is} \frac{\partial G}{\partial x_s} dt + dB(t, a).$$

As in the one-dimensional case, we "solve" these by passing to a partial differential equation (the "forward diffusion equation") in the joint probability density $\varphi(x_1, \ldots, x_r)$ of the x_i's. Using the same argument as in the one-dimensional case, we obtain

$$\frac{\partial \varphi}{\partial t} = \frac{1}{2} \sum_{i,j} \sigma_{ij}^2 \frac{\partial^2 \varphi}{\partial x_i \, \partial x_j} - \sum_i \frac{\partial}{\partial x_i} \left(\sum_s a_{is} \frac{\partial G}{\partial x_s} \varphi \right)$$

where

$$\sigma_{ij}^2 t = \int_0^1 [B_i(t, a) - B_i(0)][B_j(t, a) - B_j(0)] \, da.$$

Suppose $\partial \varphi / \partial t = 0$. Under what conditions will our diffusion equation have a solution of the form

$$\varphi = \prod_{i=1}^{r} e^{-b_i G_i}?$$

If this is indeed the case, then

$$\frac{\partial \varphi}{\partial x_i} = -b_i \frac{\partial G_i}{\partial x_i} \varphi, \qquad \frac{\partial^2 \varphi}{\partial x_i^2} = -b_i \frac{\partial^2 G}{\partial x_i^2} \varphi + \left(b_i \frac{\partial G}{\partial x_i} \right)^2 \varphi,$$

$$\frac{\partial^2 \varphi}{\partial x_i \, \partial x_j} = b_i b_j \frac{\partial G}{\partial x_i} \frac{\partial G}{\partial x_j} \varphi.$$

Substituting in our expression for φ and making use of these relations, we obtain

$$-\frac{1}{2}\sum_i \sigma_{ii}^2 b_i \frac{\partial^2 G}{\partial x_i^2} + \frac{1}{2}\sum_{i,j} \sigma_{ij}^2 b_i b_j \frac{\partial G}{\partial x_i}\frac{\partial G}{\partial x_j}$$
$$= \sum_{i,s} a_{is} \frac{\partial^2 G}{\partial x_i \partial x_s} - \sum_{i,s} b_i a_{is} \frac{\partial G_i}{\partial x_i}\frac{\partial G_s}{\partial x_s}.$$

Equating coefficients of $\partial^2 G_i/\partial x_i \partial x_s$, we obtain

$$-\tfrac{1}{2}\sigma_{ii}^2 b_i = a_{ii}, \qquad a_{is} + a_{si} = 0, \quad s \neq i.$$

Equating coefficients of $(\partial G/\partial x_i)/(\partial G/\partial x_s)$, we obtain

$$\tfrac{1}{2}(\sigma_{ij}^2 b_i b_j + \sigma_{ji}^2 b_j b_i) = -b_i a_{ij} - b_j a_j, \quad i \neq j,$$

for which the most likely solution (the only one?) is $\sigma_{ij}^2 = 0$, $i \neq j$; $b_i = 1/\theta$ for all i. Thus, if the "environmental inputs" of the different species are all independent, and if the variance σ_i^2 of each such input is equal to $-2a_{ii}\theta$, the equilibrium distribution for these equations will be the same as in the conservative case.

3.1.3 The derivatives at zero of the cross-correlations of this particular Gaussian Markov process also behave like those of the conservative theory. Let

$$r_{ij}(s) = \lim_{T\to\infty} \frac{1}{T}\int_0^T x_i(t) x_j(t+s)\, dt.$$

If we let x be the vector whose ith element is x_i, we may also write

$$r_{ij}(s) = \int x_i \varphi(x)(x_j + c_j) f(x, x+c, s)\, d^n x\, d^n c$$

where $f(x, x+c, s)$ is the probability of shifting from x to $x+c$ in the time s. This expression is

$$\int x_i x_j \varphi(x)\, d^n x \int f(x, x+c, s)\, d^n c$$
$$+ \int x_i \varphi(x)\, d^n x \int c_j f(x, x+c, s)\, d^n c$$
$$= r_{ij}(0) + \int x_i \varphi(x)\, d^n x \int c_j f(x, x+c, s)\, d^n c.$$

We are interested in

$$\left.\frac{dr_{ij}}{ds}\right|_{0+} = \lim_{\substack{\Delta s \to 0 \\ \Delta s > 0}} \left[\frac{r_{ij}(\Delta s) - r_{ij}(0)}{\Delta s}\right]$$

$$= \lim_{\Delta s \to 0} \frac{1}{\Delta s} \int x_i \varphi(\boldsymbol{x}) \, d^n x \int c_j f(\boldsymbol{x}, \boldsymbol{x} + \boldsymbol{c}, s) \, d^n c.$$

We may write

$$\int c_j f(\boldsymbol{x}, \boldsymbol{x} + \boldsymbol{c}, s) \, d^n c = \sum_{k=1}^{r} a_{jk} \frac{\partial G}{\partial x_k} \Delta s$$

and obtain, as in the Volterra theory,

$$\left.\frac{dr_{ij}(s)}{ds}\right|_{0+} = c \int x_i \sum_k a_{jk} \frac{\partial G}{\partial x_k} e^{-G/\theta} \, d^n x = a_{ji} \theta.$$

3.1.4 On the other hand, the analogy between the conservative theory and this Gaussian Markov process is by no means as complete as one might like. To establish relationships concerning community stability we must count the number of times the population curve crosses a particular level. The output of a Gaussian Markov process is very irregular if one looks at it microscopically: indeed, such a curve is nowhere differentiable. One cannot pick out the time when the curve crosses a particular line: because of its irregular motion, it will have jumped back and forth across the line infinitely often in the course of what seems to us a single crossing. Indeed, for this very reason the mathematics tells us the output curve crosses any level infinitely frequently. This peculiar mathematical artifact prevents the Gaussian Markov process from telling us anything about community stability.

3.2.1 Suppose now the environmental outputs of different species are not independent: suppose species populations are regulated by competition as well as predation.

What can we say then? Here we have to resort to linear approximations in order to proceed. We shall therefore write for

$$dx_i(t, a) = \left[\sum_s a_{is} q_s (e^{x_s} - 1)\right] dt + dB_i(t, a)$$

the linear approximation

$$dx_i(t, a) = \left[\sum_s m_{is} x_s\right] dt + dB_i(t, a).$$

If we call $m_{is} = a_{is} q_s$ component (is) of the matrix M, x_i component i of the vector X, and $B_i(t, a)$ component i of the vector $B(t, a)$, we may write our linear approximation as the vector equation

$$dX(t, a) = MX\, dt + dB(t, a).$$

3.2.2 First we shall pass to the corresponding diffusion equation to see how correlations in population fluctuations of different species relate to correlations in their environmental inputs. If

$$\sigma_{ij}^2 t = \int [B_i(t, a) - B_i(0)][B_j(t, a) - B_j(0)]\, da,$$

then the appropriate diffusion equation is

$$\frac{\partial \varphi}{\partial t} = \tfrac{1}{2} \sum \sigma_{ij}^2 \frac{\partial^2 \varphi}{\partial x_i \partial x_j} - \sum_{i,j} \frac{\partial}{\partial x_i}(m_{ij} x_j \varphi)$$

$$= \tfrac{1}{2} \sum_{i,j} \sigma_{ij}^2 \frac{\partial^2 \varphi}{\partial x_i \partial x_j} - \sum_{i,j} m_{ij} x_j \frac{\partial \varphi}{\partial x_j} - \sum m_{ii} \varphi.$$

If we assume $\partial \varphi / \partial t = 0$, $\varphi = c \exp\left(-\tfrac{1}{2} \sum_{i,j} g_{ij} x_i x_j\right)$, where g_{ij} is symmetric, then

$$\frac{\partial \varphi}{\partial x_i} = -\sum_j g_{ij} x_j \varphi,$$

$$\frac{\partial^2 \varphi}{\partial x_i \partial x_j} = \left[-g_{ij} + \sum_{l,m} g_{il} x_l x_m g_{mj}\right] \varphi$$

and the matrix g of the g_{ij} must satisfy the equation

$$-\tfrac{1}{2} \sum_{i,j} \sigma_{ij}^2 g_{ji} - \sum m_{ii} + \tfrac{1}{2} \sum_{i,j,l,m} \sigma_{ij}^2 g_{jm} x_m x_l g_{li} + \sum_{i,j,k} m_{ij} x_j x_k g_{ki} = 0.$$

We may write the above expression as

$$\sum_{i,j} \left[\tfrac{1}{2} \sigma_{ij}^2 + \sum_k m_{ik}(g^{-1})_{kj} \right] \left[\sum_{m,l} g_{jm} x_m x_l g_{li} - g_{ji} \right].$$

Since the second matrix is symmetric, the symmetric part of the first matrix must be zero, from whence we make bold to conclude

$$-\sigma_{ij}^2 = \sum_k m_{ik}(g^{-1})_{kj} + \sum_k (g_{ik})^{-1} m_{jk},$$

$(g^{-1})_{kj}$ is simply the covariance r_{kj} between x_k and x_j,

$$\frac{1}{2T} \int_{-T}^{T} x_k(t, a) x_j(t, a)\, dt = \int_0^1 x_k(t, a) x_j(t, a)\, dt.$$

This is most easily proved by noting that the average of $\sum_j g_{ij} x_j x_l$ is

$$\int x_l \sum_j g_{ij} x_j \varphi(\mathbf{x})\, d^n x = -\int x_l \frac{\partial \varphi}{\partial x_i} d^n x$$

which is equal to the unit matrix δ_{il} ($\delta_{il} = 1$ when $i = l$ and zero otherwise). Thus the average of $x_k x_j$ must be $(g^{-1})_{kj}$. Letting R be the covariance matrix for population sizes and Q be the covariance matrix σ_{ij}^2 for environmental inputs, we obtain

$$-Q = MR(0) + R(0)M^*$$

where $(M^*)_{ij} = m_{ji}$.

In the two-species case we may write

$$-\sigma_{12}^2 = m_{11} r_{12} + m_{12} r_{22} + r_{11} m_{21} + r_{12} m_{22}$$

from whence we find

$$-r_{12} = (\sigma_{12}^2 + m_{12}r_{22} + m_{21}r_{11})/(m_{11} + m_{22})$$

m_{11} and m_{22} represent crowding effects, and are therefore negative; m_{12} and m_{21} will be likewise negative if the two populations are competing for space. We find, therefore, that positive correlations in the fluctuations of the two populations may arise as a result of correlations in their environmental inputs, although they may also arise from symbiotic relationships between the populations. Negative correlations may arise if the species are competing for space. As the environmental inputs of two species are not likely to be negatively correlated, a negative correlation in their population fluctuations is good evidence for some form of competition for "space".

3.2.3 We return to our stochastic equation to calculate $R(s)$ and its derivative. We begin with

$$dX(t, a) = MX(t, a)\, dt + dB(t, a).$$

By analogy with the one-dimensional case, we multiply by a matrix function $F(t)$ and integrate to obtain

$$F(s)X(s) = F(0)X(0) + \int_0^s \frac{dF}{dt} X(t)\, dt + \int_0^s F(t)MX(t)\, dt$$
$$+ \int_0^s F(t)\, dB(t, a).$$

We must therefore solve the equation

$$dF/dt = -FM, \qquad dF^*/dt = -M^*F^*.$$

If $F(0) = F^*(0)$ is the unit matrix, then

$$F^*(t) = e^{-M^*t}, \qquad F(t) = e^{-Mt},$$

and we accordingly conclude

$$X(s) = e^{Ms}X(0) + e^{Ms}\int_0^s e^{-Mt}\, dB(t, a).$$

3.2.4 If s is positive we may write

$$r_{ij}(s) = \lim_{T\to\infty} \frac{1}{2T} \int_{-T}^{T} x_i(t) x_j(t+s)\, dt$$

$$= \int \left[\sum_k (e^{Ms})_{jk} x_k \right] x_i \varphi(x)\, d^n x$$

$$= \sum_k (e^{Ms})_{jk} r_{ki}(0),$$

from whence we make bold to conclude

$$R(s) = R(0) e^{M^*s}, \qquad \lim_{s\to 0^+} \frac{dR}{ds} = RM^*.$$

When $R_{ij}(0) = 0$ for $i \neq j$, we find

$$\lim_{s\to 0^+} \frac{dr_{ij}(s)}{ds} = m_{ji} r_{ii} = q_j a_{ji} r_{ii}$$

in analogy to the Volterra case. I suspect that this rule applies to quite a wide range of stochastic equations, and that the derivatives of these correlations, like Liapunov's stability criteria, depend only on the linear approximation to the process concerned. Such a result could be formulated as the following:

UNPROVED ASSERTION. Let $dX(t) = G(X)\, dt + dB(t)$ be a stochastic equation. Let $G(0) = 0$. (This is no real restriction, since if $G(\bar X) = 0$ we can rewrite our equation in terms of $X - \bar X$.) Let $-X \cdot G(X)$ be "positive definite". Then

$$\lim_{s\to 0^+} \frac{dR}{ds} = RM^*,$$

where

$$m_{ij} = (\partial g_i / \partial x_j)|_{x=0}.$$

REFERENCES: PART III

Much of this chapter is a summary of the section on (linear) "Gaussian Markov Processes" in

S. R. de Groot and P. Mazur, *Nonequilibrium thermodynamics*, North-Holland, Amsterdam, 1962.

IV. Numerical Examples, Leading to an Analysis of the Oscillation of Lynx and Hare in Canada

4.1.1 In the last chapter we developed a body of theory enabling us to infer the nature and strength of species interactions from characteristics of their population curves. Now we shall illustrate the uses of this theory by applying it to a pair of numerical examples, one an experimental situation where we already know the interactions, and one a natural oscillation whose causes are quite mysterious.

4.2.1 To check our method, it seems advisable to analyse an oscillation whose causes we already know. The best data available are those from an experiment by Huffaker [1] in which a predator-prey oscillation was created in the laboratory. Huffaker used as prey a mite feeding on oranges, and employed another species of mite as his predator. By arranging pieces of orange in a suitably complicated way, Huffaker made it difficult for the predator to find and extinguish all the prey: he thus managed to create an oscillation executing three cycles in seven months before the predators died out. He sampled his populations every five days, forty readings in all. He presents his data in a graph, from which I have read off the numbers given in Table 1.

4.1.3 We shall perform our analysis in terms of the linear approximation

$$dx_1(t) = m_{11}x_1\,dt + m_{12}x_2\,dt + dB_1(t),$$
$$dx_2(t) = m_{21}x_1\,dt + m_{22}x_2\,dt + dB_2(t),$$

where $x_1(t) = \log N_1(t) - \overline{\log N_1(t)}$ is the difference between the logarithm of the number of prey at time t and its mean: $x_2(t)$ carries a similar meaning for the predators.

We shall calculate the m_{ij} from the equation

$$\lim_{s \to 0^+} \frac{dR(s)}{ds} = R(0)M^*$$

which, in the two-species case, gives us the equations

$$(dr_{11}/ds)|_{0^+} = r_{11}(0)m_{11} + r_{12}(0)m_{12},$$
$$(dr_{21}/ds)|_{0^+} = r_{21}(0)m_{11} + r_{22}(0)m_{12},$$
$$(dr_{12}/ds)|_{0^+} = r_{11}(0)m_{21} + r_{12}(0)m_{22},$$
$$(dr_{22}/ds)|_{0^+} = r_{21}(0)m_{21} + r_{22}(0)m_{22}.$$

The r_{ij} are defined by the equation

$$r_{ij}(s) = \lim_{T \to \infty} \frac{1}{T} \int_0^T x_i(t)x_j(t+s)\,dt.$$

We shall estimate the $r_{ij}(s)$ by averages of $x_i(t)x_j(t+s)$ extended over the times the populations were sampled: we have tabulated appropriate values in Table 2 [2]. In particular,

$$r_{11}(0) = .11, \quad r_{22}(0) = .27, \quad r_{12}(0) = r_{21}(0) = .007 \cong 0.$$

We shall write

$$\lim_{s \to 0^+} \frac{dr_{ij}(s)}{ds} = \tfrac{1}{5}[r_{ij}(5) - r_{ij}(0)]$$

where the time is measured in days. We accordingly obtain

$$(dr_{11}/ds)|_{0^+} = -.004 \qquad (dr_{12}/ds)|_{0^+} = .008$$
$$(dr_{22}/ds)|_{0^+} = -.01 \qquad (dr_{21}/ds)|_{0^+} = -.014$$

from whence we make bold to conclude

$$m_{21} = \frac{1}{r_{11}} \frac{dr_{12}}{ds}\bigg|_{0^+} \cong .07 \qquad m_{11} = \frac{1}{r_{11}} \frac{dr_{11}}{ds}\bigg|_{0^+} \cong -.036$$

$$m_{12} = \frac{1}{r_{22}} \frac{dr_{21}}{ds}\bigg|_{0^+} \cong -.05 \qquad m_{22} = \frac{1}{r_{22}} \frac{dr_{22}}{ds}\bigg|_{0^+} \cong -04.$$

We may calculate the natural frequency of oscillation of this system by finding the imaginary component of the eigenvalues of the matrix M: this is roughly .06. The period of oscillation of this system should accordingly be $2\pi/.06 \cong 100$ days, which compares not unfavorably with the observed 70. Our analysis brings out the importance of "crowding effects": $|m_{11}|$ and $|m_{22}|$ are not far smaller than $|m_{21}|$ and $|m_{12}|$. Presumably the chance fluctuations of small numbers counterbalance the damping effect so introduced.

4.2.1 Now we turn to an example from nature. For some years, the Hudson's Bay Company has trapped the Canadian lynx and its primary food, the varying hare. They have caught what they could, and kept yearly records of the numbers caught. However inaccurately these catches may reflect events in the wild, the Company records still represent the best information available on long-term population changes.

These data reveal remarkable fluctuations in population size, whose causes have been the subject of considerable controversy among biologists. Although some have sought to deny this, the fluctuations are clearly periodic. If one considers the autocorrelation function

$$\tilde{r}(s) = \frac{\frac{1}{50}\sum_{t=1847}^{1897} \log N(t) \log N(t+s) - (\frac{1}{50})^2 \sum_{1847}^{1897} \log N(t) \sum_{1847+s}^{1897+s} \log N(t)}{\frac{1}{50}\sum [\log N(t) - \overline{\log N(t)}]^2}$$

for the logarithms of lynx catches from 1847 to 1903, one finds that $\tilde{r}(0) = 1$, $\tilde{r}(4) = -.7$, $\tilde{r}(8) = .4$. The lynx population is negatively correlated with its value four years earlier, and positively correlated with its value eight years earlier: we clearly have to do with an eight-year cycle. For the hares, $\tilde{r}(0) = 1$, $\tilde{r}(5) = -.5$, $r(10) = .3$: the period of the oscillation is longer, and the periodicity is less strongly marked, but still definite. As the autocorrelations would suggest, the lynxes exhibit six well-marked population maxima during this fifty-seven year period, while the hares exhibit only five.

What gives rise to these oscillations? It is appealing to believe that they represent a natural predator-prey oscillation, and we shall apply the techniques of the preceding section to find out whether this is indeed so. The data I shall use are taken from a work by MacLulich [3] on the varying hare: as my numbers were read from a rather small graph, I have tabulated them (Table 3) so that the reader may check them against the original if he wishes. I have focussed on events between 1847 and 1903, as the interpretation of the record for these years involves the fewest difficulties.

4.2.2 Once again, as in 4.1.3, we determine the m_{ij} from the equation

$$\lim_{s \to 0^+} \frac{dR(s)}{ds} = R(0)M^*.$$

In Table 4 we list values of $r_{ij}(s)$: for s less than 8 they were calculated by the rule

$$r_{ij}(s) = \tfrac{1}{50} \sum_{t=1847}^{1897} x_i(t)x_j(t+s)$$

where $x_i(t) = \log N_i(t) - \overline{\log N_i(t)}$. For $s > 7$, we extend our average over as long a period as possible. We set

$$\lim_{s \to 0^+} \frac{dr_{ij}(s)}{ds} = r_{ij}(1) - r_{ij}(0).$$

We accordingly obtain

$-.07 = .19m_{11} + .06m_{12}, \quad -.05 = .06m_{11} + .16m_{12}$

from whence we find $m_{11} = -.3$, $m_{12} = -.2$: we also obtain

$.04 = .19m_{21} + .06m_{22}, \quad -.05 = .06m_{21} + .16m_{22},$

from whence we find $m_{21} = \frac{4}{11}$, $m_{22} = \frac{7}{16}$. Crowding effects are relatively more important here than in Huffaker's experimental system. The imaginary part of the eigenvalues of M is nearly $\frac{1}{4}$, from whence we find that the system should oscillate with a period of 25 years. In fact it has a period of ten years: the oscillation accordingly must have a different cause. And indeed, Dr. MacArthur informs me (personal communication) that the hares oscillate on islands where lynxes are absent: the oscillation may involve the grass they eat, or some disease.

It is appealing to believe that the lynxes follow a hare oscillation that is caused by some other mechanism. But the lynxes are most closely correlated with the abundance of hares a year earlier. This is a most extraordinary circumstance in view of the fact that, left to themselves, lynxes and hares would oscillate every 25 years (such a fact suggests a four to six-year timelag in the response of lynx population to food supply, rather than the year lag actually observed.) Dr. MacArthur informs me that strong crowding effects would serve to shorten the lag, and indeed these are present here. But I have not calculated the extent to which they would reduce the lag. I find the whole oscillation most mysterious.

TABLE I: ABUNDANCE OF PREY AND PREDATOR IN HUFFAKER'S EXPERIMENT

Day	No. of Prey $[N_1(t)]$	$\log N_1(t)$	No. of Predator $[N_2(t)]$	$\log N_2(t)$
5	160	2.20	1	0
10	600	2.78	3	.48
15	1000	3.00	12	1.08
20	1400	3.15	28	1.45
25	900	2.95	36	1.56
30	500	2.70	25	1.40
35	250	2.40	17	1.23
40	200	2.30	12	1.08
45	150	2.18	8	.90
50	200	2.30	7	.85
55	200	2.30	6	.78
60	250	2.40	3	.48
65	350	2.54	1	0
70	450	2.65	1	0
75	500	2.70	3	.48
80	850	2.93	5	.70
85	1600	3.20	5	.70
90	1750	3.24	15	1.18
95	2000	3.30	19	1.28
100	2000	3.30	20	1.30
105	1700	3.23	16	1.20
110	1400	3.15	15	1.18
115	800	2.90	35	1.54
120	500	2.70	18	1.26
125	550	2.74	7	.85
130	650	2.81	7	.85
135	700	2.85	12	1.08
140	700	2.85	6	.70
145	850	2.93	1	0
150	1150	3.06	0	0
155	1800	3.26	1	0
160	1900	3.28	1	0
165	1800	3.26	1	0
170	1700	3.23	3	.48
175	1800	3.26	9	.95
180	1700	3.23	20	1.30
185	1500	3.18	30	1.48
190	500	2.70	39	1.69
195	200	2.30	25	1.40
200	100	2.00	8	.90

Table II: Covariance in Fluctuations of Experimental Predator and Prey Populations (Measured Logarithmically)

	r_{11}	r_{12}	r_{21}	r_{22}
$s=0$	.13	.007	.007	.27
$s=1$	.11	.047	−.061	.22
$s=2$		.069	−.107	

Table III: Abundance of Lynx and Hare in Canada

Year	No. of Hares $[N_1(t)]$	$\log N_1(t)$	No. of Lynxes $[N_2(t)]$	$\log N_2(t)$
1847	21,000	4.32	49,000	4.70
1848	12,000	4.08	21,000	4.32
1849	24,000	4.38	9,000	3.95
1850	50,000	4.69	7,000	3.85
1851	80,000	4.90	5,000	3.70
1852	80,000	4.90	5,000	3.70
1853	90,000	4.95	11,000	4.04
1854	69,000	4.84	22,000	4.34
1855	80,000	4.93	33,000	4.52
1856	93,000	4.97	33,000	4.52
1857	72,000	4.86	27,000	4.43
1858	27,000	4.43	18,000	4.26
1859	14,000	4.15	8,000	3.90
1860	16,000	4.20	4,000	3.60
1861	38,000	4.58	4,000	3.60
1862	5,000	3.70	4,000	3.60
1863	153,000	5.18	20,000	4.30
1864	145,000	5.16	35,000	4.54
1865	106,000	5.02	68,000	4.83
1866	46,000	4.66	70,000	4.85
1867	23,000	4.32	40,000	4.60
1868	2,000	3.30	22,000	4.34
1869	4,000	3.60	9,000	3.95
1870	8,000	3.90	5,000	3.70
1871	7,000	3.85	4,000	3.60
1872	60,000	4.78	10,000	4.00
1873	46,000	4.68	18,000	4.26
1874	50,000	4.70	19,000	4.28
1875	103,000	5.01	43,000	4.63
1876	87,000	4.94	37,000	4.57
1877	68,000	4.83	22,000	4.34
1878	17,000	4.23	15,000	4.18
1879	10,000	4.00	10,000	4.00

TABLE III, CONT.

Year	No. of Hares $[N_1(t)]$	$\log N_1(t)$	No. of Lynxes $[N_2(t)]$	$\log N_2(t)$
1880	17,000	4.23	8,000	3.90
1881	16,000	4.20	8,000	3.90
1882	15,000	4.18	30,000	4.48
1883	46,000	4.66	52,000	4.72
1884	55,000	4.74	75,000	4.88
1885	137,000	5.14	80,000	4.90
1886	137,000	5.14	33,000	4.52
1887	95,000	4.98	20,000	4.30
1888	37,000	4.57	13,000	4.11
1889	22,000	4.34	7,000	3.85
1890	50,000	4.70	6,000	4.78
1891	54,000	4.73	10,000	4.00
1892	65,000	4.81	20,000	4.50
1893	60,000	4.78	35,000	4.54
1894	81,000	4.91	55,000	4.74
1895	95,000	4.98	40,000	4.60
1896	56,000	4.75	28,000	4.45
1897	18,000	4.28	16,000	4.20
1898	5,000	3.70	5,000	3.70
1899	2,000	3.30	6,000	3.78
1900	15,000	4.18	10,000	4.00
1901	2,000	3.30	21,000	4.32
1902	6,000	3.78	35,000	4.54
1903	45,000	4.65	50,000	4.70

TABLE IV: COVARIANCE IN POPULATION FLUCTUATIONS OF LYNX AND HARE (MEASURED LOGARITHMICALLY)

	r_{11}	r_{12}	r_{21}	r_{22}
$s = 0$	.18	.06	.06	.16
$s = 1$	.11	.10	.01	.11
$s = 2$	.02	.05	−.04	.04
$s = 3$	−.03	.03	−.09	−.04
$s = 4$	−.09	−.05	−.11	−.11
$s = 5$	−.11	−.08	−.08	−.12
$s = 6$	−.08	−.09	−.04	−.08
$s = 7$	−.07	−.06	.02	−.0
$s = 8$	.02	−.02	.11	.07
$s = 9$	.06	.04	.11	
$s = 10$	.07		.10	
$s = 11$	.08			

References: Part IV

1. C. B. Huffaker, *Experimental studies on predation: Dispersion factors and predator-prey oscillations*, Hilgardia **27** (1957), 343–383.

2. We have tabulated correlations of logarithms taken to the base 10: theory concerns itself, as always, with logarithms to the base e. Taking logarithms to the wrong base merely introduces a scale factor. As we are only concerned with expressions of the form $m_{ij} = (1/r_{jj})(dr_{ji}/ds)$ we find that the scale factors will cancel out, and that we have not wandered into error.

3. D. A. MacLulich, *Fluctuation in the number of the varying hare*, Univ. of Toronto Press, Toronto, Ontario, Canada, 1937.

PRINCETON UNIVERSITY

Evolution of Complex Genetic Systems

Richard C. Lewontin

Having learned that MacArthur and Levins were not going to be present, and feeling that, for an audience of mathematicians, a discussion of complex genetic systems in a vacuum would not be particularly useful, I want to put the topic in a more general framework which seems to me a reasonable one for people interested in mathematical biology. That is to say, I want to start by stating what I conceive to be the problems or the subproblems of the field of population biology insofar as they have mathematical interest, although that mathematical interest may be trivial from the mathematicians' standpoint. Then I want to illustrate these problems with some more technical and exact material from the talk that I would have given had my confreres been here.

I will start out by saying that what is done in the field of mathematical biology, at least in the field of population biology, is mathematically trivial. The repertoire of techniques and concepts required by the biologist, in his present state of sophistication about biology, are concepts and techniques that to a mathematician are not worth giving a second thought. They are concepts and techniques which I would describe as manipulative: how one manipulates a matrix equation, how one solves some differential equations, or, if one can't solve them, how one gets numerical approximations, something about stochastic processes, the kind of applied mathematics that the field of mathematics has by and large left behind. I don't mean to say that there aren't valid fields of mathematical inquiry in the fields of stochastic processes and differential equations, for example, but they are at a level which is far beyond the problems of the biologist. Most biologists, for example, are not interested in problems of existence of solutions—they want the numbers, and this often drives them to deal with the problem strictly in a manipulative way.

Because the biologist knows so little about modern mathematics, he conceives his problems in a framework of classical, simple applied mathematical techniques, and if there is to be any fruitful cooperation between mathematicians and biologists, it is that the mathematicians will have to see, in our description of our problems, realms of mathematical technique which we don't even know exist. We don't need mathematicians to help us solve differential equations.

With that preface, let me say what I consider to be the theoretical problems, and in fact the general problems, of population biology. Population biology has as its *raison d'etre* the explanation of why there are the numbers and kinds of animals and plants that there are at a particular place in the world, and why there used to be different numbers and kinds, and what the numbers and kinds are likely to be in the future at some specified time. That is, population biology is concerned with a causative explanation for the variety of organisms in space-time, and all population biology, however one attacks it, must necessarily be concerned with that problem. There are people who call themselves physiological ecologists, for example, who are concerned with the physiological processes that plants and animals exhibit in a particular place given the water table, the light, etc. but the reason that they are interested in these questions is not simply because they are physiologists, but because they believe that the answers will help them answer the main question: Why is there an oak tree growing in this place, and why is there not an oak tree growing in that place?

Given this general question, which includes the genetical questions of why there is a *specific* kind of oak tree growing here and not there, and why it used to be that there weren't any oak trees, and how long will oak trees be around in the evolutionary future, there are two kinds of explanations that have to be blended in the population biologist's thinking. One I would classify as essentially an historical explanation, and the other might be called determinate rather than deterministic. To distinguish between these, I mean the following: To what extent is the current distribution of animals and plants on the surface of the earth a result of unique historical processes which represent only a sample from a universe of possible outcomes, given the same physical environment in which the earth finds itself? Or, to what extent is the present situation, as far as the distribution of kinds and numbers of animals and plants are concerned, the result of forces that have impinged on the system irrespective of the starting point? To put this in terms which are a little more exact and perhaps a little more comfortable to the applied mathematician, imagine that we have a potential surface in n dimensions. The question is, how many singularities are there on this surface? How many of these singularities are stable, and is any particular population sitting at a singularity or in a singular region (in a potential well, if you like) because this is the only potential well to which it might have gone? Or is it that there is a multiplicity of potential wells and the one to which the population has gone is the result of the unique solution of the particular differential equation involved?

Superimposed on this problem, of course, is the problem of how exactly we wish to specify the location of these potential wells or singularities. After all, no population is exactly stable in its configuration. The numbers of the different kinds of organisms are fluctuating in time at any particular place, and when we describe a population as having a stable configuration, what we mean is that there is a region in the space of n dimensions, where the population is found with high probability. In fact, the best analogy is an electron cloud, a cloud of potential. The population has a certain probability of finding itself in some cell in the phase-space at every moment in time. We wish to describe the probability of the cells in the phase-space, and ask the question of whether the local region in which the population is located is the only region available in the space. From an evolutionary standpoint, especially from a genetical standpoint, we know something about the answer to this problem. We know the outcome of some evolutionary processes is opportunistic, that there exist multiplicities, only one of which has been chosen. From an ecological standpoint, however, for the question, why is there an oak tree here and not an oak tree there, we don't really know the answer. It is not at all clear whether, given a small number of physical facts about the environment, I could predict, within an acceptable error, the probability distribution of the different species that occupy a region. There is considerable argument and discussion among ecologists about the extent to which this is true. There is secondly the problem, even among those who admit that it is true, of how many different factors have to be brought into play in order to make a proper explanatory hypothesis.

This brings me to a last point which is at issue among population biologists, and that is the relative role of main effects and interactions in determining the face of the earth. Every biologist knows, at one level of sophistication, that the world is unique, and that the complete explanation of the location of every animal and plant requires a complete specification of all the physical and biotic factors in the world. That is to say, when we blink, it is felt on the farthest star. Yet no one, even those involved in space travel, is very much concerned with the gravitational effect on a space ship of my blinking. We ignore this kind of interaction. In the same way, the biologist, as a matter of practical fact, not as a matter of *a priori* model making, must discover to what extent the configuration of living organisms in the space-time continuum can be predicted and explained by a relatively small number of factors acting essentially as main effects and to what extent he has to take into account the unique interactions. The answer in a trivial sense is clear: there will be some cases where interaction is very important and some cases where it is not. But in a global sense, we would like to be able to say that for most organisms and for most communities it will be possible from five or six things about the physical environment, including perhaps its characteristics in time, to predict what kinds of organisms there will be.

What MacArthur would have talked about is one element of this problem: why are there 43 species of birds on a given island, and what is there about the distance of the island from the mainland, the size of the island and the number of species in the mainland community which can help account for the fact that there are precisely 43 species? Moreover, what is the distribution over time of the number of species on this island? He would have discussed an equilibrium stochastic theory of island population of the following nature· an island receives immigrants from the mainland at a certain rate, this rate proportional to the angle ubtended by the island when looked at from the mainland, since the birds presumably fly more or less at random. It is also proportional to the actual distance from the mainland. The birds that land on this island have a certain probability of survival and reproduction on the island, which probability decreases as the number of resident species increases—it gets harder and harder for the next ones to get in, since some of the resources are being used. Finally, there is a certain rate of extinction of species. The probability distribution of the number of species on an island will be given by the intersection of the extinction curves and the immigration curves, these being in turn determined by the size of the island. This is what is called an equilibrium theory of insular zoogeography, and the real question at issue is whether one can, from a small number of factors like size of the island, distance of the island from the mainland, something about the complexity of the vegetation on the island, and something simple about the numbers and kinds of species on the mainland, predict the average species composition of such an island. He would also have discussed the quantitative predictions made on the basis of a little simple stochastic theory—the numbers of species on the islands performing a random walk—and he would have compared this prediction with some actual observations relating the size of the islands, their distance from the nearest island, and from the mainland. This is the kind of model-making which goes on in population biology. Whether or not this particular model of insular zoo-geography is correct, everyone who tries to explain the numbers of species on an island does essentially this.

Superimposed on this question of describing the current state and the equilibrium situation and the dynamics of reaching that state is a second problem that has recently become interesting to some of us, about which Levins would have talked. This is the notion that there may be a difference between the equilibrium situation or the dynamic process reaching equilibrium, on the one hand, and some criterion of optimality on the other. That is to say, one can, by some *a priori* reasoning, arrive at a different sort of surface than the potential surface that I just spoke about. The potential surface is the actual surface created by the dynamics of the biological process. But one can create a surface of utilities, to the world, to the species, to God, to man, to whatever it is that creates these utilities, and say that it really would be better if there were an oak tree here and not there. One of the problems for people in population biology who want to create a set of utilities is that there is obviously no universal agreement on how one goes about defining utility. There are various ways in which it has been done, but I think it's clear to you that, depending on the definition of utility, the potential surface about which I just spoke, given by the kinetics of the process, and the utility surface may or may not coincide, and the degree of their coincidence will be almost an accident of how you chose the utilities to begin with. Nevertheless, there is a great deal of attention currently paid in population biology to the concept of optimal behavior of populations, optimal states of populations, evolution as an optimizing process, whether we mean genetical evolution or evolution of biological communities. It is perhaps here that I differ most from my two previous colleagues who have just spoken. I began some years ago by believing that the principle of optimality would be a very useful one in biology and I am actually guilty of having written a paper proposing it. I have since become somewhat disenchanted with the principle of optimality in biology, because I find that it is in a sense epiphenomenal. It is not very interesting to know that a state is optimal if the biological system cannot, by its own dynamic, reach that state. It is not terribly interesting to know that the stable equilibrium point given by the dynamics of an evolutionary process happens to be different from the optimal point defined by some criterion of optimality. On the other hand, suppose they do coincide, or nearly so. Then knowing that the particular point is a point of optimality doesn't help you in any predictive sense. Therefore it isn't entirely clear to me why the principle of optimality is useful in biology. Nevertheless, in order to be perfectly fair and unbiased, I have to say that it is not clear that it is *not* useful, and that there is a whole series of mathematical problems associated with optimization. These are problems which you are all, as mathematicians and as applied mathematicians, well aware of. May I say that the problems of optimization in population biology pose very particular difficulties for mathematics, because most of the assumptions of, say, game theory and utility theory, as derived by sociologists and econometricians, do not apply even roughly to biological systems. Let me mention two of them which seem to me of very great importance to the biologist. One is the principle of transitivity. If you read an elementary book on game theory, say Luce and Raiffa, a great deal is made of transitivity. If I prefer A to B and B to C, then rationally I will prefer A to C. But biological organisms don't always behave in this way, alas; there are significant departures. It is possible to show experimentally that if, in a competitive situation, genotype A eliminates genotype B, and in another situation genotype B eliminates genotype C, it does not follow at all that A will eliminate C in direct competition. On the contrary, there may be a unique interaction which reverses this. Nor is this phenomenon particularly rare. Thus the principle of transitivity of utilities is not necessarily true in biology, and we must build a theory of utility functions that allows for nontransitivity.

Second, the notion of the linear utility function is not applicable in biology. It is not true that you can find out the utility of a mixed strategy by drawing a straight line between pure strategies in a Cartesian space. In fact, the question of what the form of the geodesic is for mixed strategies is one that has no general solution: it has to be discovered in each experimental situation. The only thing I am certain of is that it is almost never linear. That is because a mixture of organisms refuses to behave as an additive mixture of independent elements, and the moment the mixture of organisms takes on properties of its own, then clearly the line connecting the utilities is not a straight line, at least not in an ordinary Cartesian space. Knowing nothing about modern developments in game theory, I can't say whether these problems have been solved or not. I propose that they are problems which need very careful attention—nonlinearity of utility functions, and nontransitivity of utilities—if this kind of approach is to be useful to the biologist. To put it in another context, if we are going to use, let's say, the simple theory of convex sets, to solve the optimality problem, we have to do it in a space that is not Cartesian.

Now, I want to illustrate some of the specific problems that I myself have been concerned with, and they will illustrate, essentially, the first statement I made about the problems of population biology. I will talk about investigations of the equilibrium genetical configurations of populations, attempt to predict what the equilibrium configurations are, and how the equilibrium configurations will change when the parameters of the models change. Second, I will discuss the attempt to describe the actual dynamics in time of the process which may or may not be going to an interesting equilibrium. It is not true that all biological processes go to an interesting equilibrium; many go to equilibria that are trivial in the biological sense. They go from a state of heterogeneity to a state of homogeneity and biologists are not very much interested in a homogeneous state. I will describe the equilibria of heterogeneous systems and how one goes about investigating the dynamics by which the population reaches these states. In particular, I will not talk about an ecological problem, which is: given, that I have a field of land with no organisms in it, or no animals, or no trees, what is the history of the occupation of this plot of land by trees and animals? I will talk rather about a genetical problem, which is: given that different genetically determined types in a population have different probabilities of survival and reproduction, what are the changes that occur in the relative frequencies of the different types in the population?

Evolutionary genetics is in a unique status in population biology, and in biology as a whole, in that it is not required of the mathematician that he help to create the basic model. We are not in the business of fitting a number of observations to an unknown model; we don't have a black box, whose structure we are trying to discover. The geneticist knows exactly what the gears inside the black box look like, and that is why, from the mathematical standpoint, his problem is manipulative only. The problem of the population geneticist is: if a force is applied to the mechanism at one point, what will be the output at another point? This can be only a manipulative problem of applied mathematics. In fact, one can avoid mathematics entirely by creating a simulation of the problem, putting input into one end and getting the answers out of the other. A great deal of simulation is in fact done. However, as an aesthetic question, we are interested in trying to make more general statements about the relations between inputs and outputs than the simple observation of the correspondence between them in a particular case. I will discuss some of the more general problems of relating the outputs to the inputs in both qualitative and quantitative terms.

Mendelism has given to the population geneticist the mechanism by which his system works, and all of mathematical genetics is nothing but an elaboration of the laws of Mendel, plus the laws of recombination, which are a kind of afterthought and which are equally mechanical. There is nothing in population genetics beyond Mendelism, recombination, and the fact that different genotypes leave different numbers of offspring. That, of course, is a critical point, but it is also a fairly obvious one.

Assume a population made up of organisms, each with a large number of genes which control the physiological and behavioral aspect of the organism. Each one of these genes is capable of existing in a number of different states, called alleles. The first problem for the population geneticist is, what is the dimensionality of the space required to make an adequate description of this population? Let us suppose that there are L of these loci, and that at each locus there could be any one of a alternative configurations. Then there are clearly a^L possible types in the population. But that does not mean that we need such a dimensionality to make an adequate description of the population. For example, let $a = 2$. The two alleles have frequencies of P_L and $1 - P_L$, and therefore at each locus, a single parameter is sufficient to describe the state of that locus in the population. Then, clearly, since there is only one parameter required for each locus, L parameters would be sufficient to describe the whole population. It might be sufficient to know the allele frequency at every locus. We have a hypercube of L dimensions, and the description of the population is given by a point somewhere in this unit hypercube. But the alternative possibility is that we must know the *joint distribution* of all the alleles at all loci, that it is not sufficient to know the gene frequency P_L at the Lth locus for all L's. In this case we must specify the distribution of all possible combinations in the population. There are 2^L of these, and the description of the population is a point in a hypertetrahedron of dimensionality $2^L - 1$. It is a hypertetrahedron because the dimensions are not independent. The sum of all frequencies must be unity and no frequency can fall outside the interval $[0, 1]$.

If we wish to predict the equilibrium composition of a population, which of these dimensionalities do we need? The answer depends on the degree to which alleles at different loci interact in determining morphology and physiology. If I know the contribution to the organism of the substitution of a 0 for a 1 at one locus, irrespective of the state of other loci, L dimensions is sufficient. But if the effect of such a substitution depends upon the state of another locus, then we need the higher dimensionality.

For many years, it has been assumed, in considering the changes that occur in populations and in predicting the equilibrium configuration of populations, that the lower dimensionality is sufficient. It seemed reasonable that, even though there were interactions between loci in determining the characteristics of individuals, this would iron out if you gave it enough time, and that the equilibrium situation in a population would be the same, irrespective of whether there were such interactions, because of recombinations between chromosomes. If there is a chromosome with configuration 001100111010, and another one which is 110011000101, the process of recombination of these gene strings to produce new combinations can generate all possible combinations such as 111111111111. If this goes on long enough, then all loci will become randomized with respect to each other and it will not matter whether there is any interaction. We could then say that even though in theory the dynamics of the process might require the description of the motion of a point in a 2^L-dimensional tetrahedron, at equilibrium, it was sufficient to specify the situation by L dimensions. When we look at this process more closely, however, we find that this is not true. There is a great variety of conditions under which this is not a sufficient description of the population; that in fact, many populations may have identical configurations in L-dimensional space, but quite different configurations in 2^L-dimensional space. If this were just a matter of description, then as biologists, we would not care. But it is important to be able to distinguish between the points in 2^L-dimensional space and their reflection or projection in the L-dimensional space because it makes a difference to what happens in evolution.

I would now like to describe how one handles this problem, and to do so I will introduce the notion of the genetic potential function of a population. Each genetical type in a population has a certain probability that an egg of that type at a given time will be represented by a progeny egg at some future time. Assume discrete generations; let's say a grasshopper, that has only one generation a year, so that we really know what we mean by the frequency of different kinds of genes in different generations. Each egg laid by a grasshopper in 1965 has some probability of becoming an adult grasshopper ready to lay an egg in 1966, and if it is a grasshopper ready to lay an egg, there is some probability distribution of the number of eggs it will lay. This can be put together to say that what we are really concerned with is the probability that an egg of a given type is represented by a progeny egg in the next generation. Let us call these different probabilities the "fitnesses" of the different genetic types. Let W_i be the fitness value for the ith diploid genotype; we define $\bar{W}$ as the ordinary arithmetic mean, the weighted mean fitness of the population. $\bar{W} = \sum Z_i W_i$. Z_i, in turn, is a function of the P_j, the frequency of one of the alternative allelic genes at the jth locus. Given a reasonably simple genetical situation, and if we are in the L-dimensional space, where we can regard each gene-locus as being independent, then we can write the following equation:

$$\Delta P_i = P_i(1 - P_i)\partial \bar{W}/\partial P_i.$$

That is to say, in a very restricted sense, the rate at which the frequency of a given genetical type changes in the population is proportional to the product of a weighting function times the partial of $\bar{W}$ with respect to P_i. This immediately tells us that $\bar{W}$ is a potential function, and that when this potential function has a singularity, ΔP_i is zero. This means that we can predict not only how the genetical composition of the population changes, but if we just describe singularities on the $\bar{W}$ surface, then we know what all the equilibrium situations are. Moreover, we can describe whether they are stable or unstable, whether they represent minimaxes in the L dimensions, and so on, by the usual methods of simple problems of maximization in L dimensions. Then this $\bar{W}$ surface, if everything is very simple, is a predictor of the equilibrium situation in the population. Notice that the dynamics of change of the genotype is not the simple motion on the potential surface the way the classical particle is, because it has the weighting function $P_i(1 - P_i)$. It says that types that are rare move slowly, and that types that are frequent move faster. One of the mathematical problems that exists is how many singularities of each type, stable, unstable, and metastable, exist strictly inside the hypercube given the form of the function $\bar{W}$. Now, $\bar{W}$ usually has a fairly simple form, under the simplest conditions, that is to say, it is a product of L quadratics. The mathematical problem is: what can I tell about the number of singularities strictly inside the unit hypercube on which it is defined, if I know the values of the individual W's? It is not a trivial problem—I know the answer for one, two, and I think I know the answer for three dimensions. I may know the answer for N dimensions, but I can't prove it. This problem has not been solved at the present time. Why is it interesting to know the number of singularities strictly inside the open region (0, 1)? If there is a stable point anywhere on one of the $(L - 1)$-dimensional cells, edges, or vertices, it means that one of the types is missing. This reduces the dimensionality of the problem by one, and we are again back inside a hypercube of a lower dimension. Secondly, we care how many singularities there are because this tells us how many paths evolution of this population could take, and the locations of these tell us something about where the population will go from any given initial conditions. But we must know more about these than whether they represent stable or unstable equilibria, whether they are stable in the sense of neighborhood stability. By neighborhood stability, I mean that the system returns from an infinitesimal perturbation and this can be tested by the Lyapounov criterion. But neighborhood stability in this sense is not really very interesting to the biologist, because the perturbations that occur in a biological system are never arbitrarily small. They are often very large and moreover, we are not dealing with a space-continuous situation. If the population consists of a finite number of individuals, then P_i can take only certain rational values. This means that every change has a finite step size. With finite step size and discrete time, what do we mean by a stable equilibrium point? We must think of these stable regions as potential wells of finite size. What is the area of the hypersurface that a particular well occupies? How far can the point move and find itself in the influence of a different potential well? Moreover, for a perturbation of a given size, how strong is the restoring force? For example, it is important to know the difference between a potential well of shallow slope but large area, and a potential well of steep slope but small area. In nature, the population's configuration is being constantly perturbed by random elements, by changes in the environment, by sampling errors from one generation to the other. As a result, the population in any one generation will take a step of some size, in some direction in the hypercube, and there will be a restoring force from that point. Over time, the population will follow some kind of a path. What is the probability density in time of the location of the population in the space? What difference do the two types of potential well make to this probability density? For perturbations of a given size or distribution of sizes, which type of potential well has a greater "stability"?

The general problem of stability becomes vastly worse if I have to go from the L-dimensional space to the $(2^L - 1)$-dimensional tetrahedron, because a single stable state in the cube may correspond to a number of alternative stable states in the tetrahedron. I want to illustrate some of the complexities of the configurations of these higher dimensional surfaces: some are not everywhere differentiable—they have creases in them—the potential surfaces sometimes degenerate catastrophically with a very small change in parameter values, like a wave that breaks. A breaking wave is Rene Thom's example of structural instability, and these genetic systems have such structural instabilities. I am going to illustrate how there are critical parameter values, especially of the amount of recombination between genotypes, such that on one side of such a critical parameter value, there is a simple surface and everything is predictable by the simple equations, while on the other side of that critical value, the surface breaks up into pieces and there are a number of possible configurations that the population may go to.

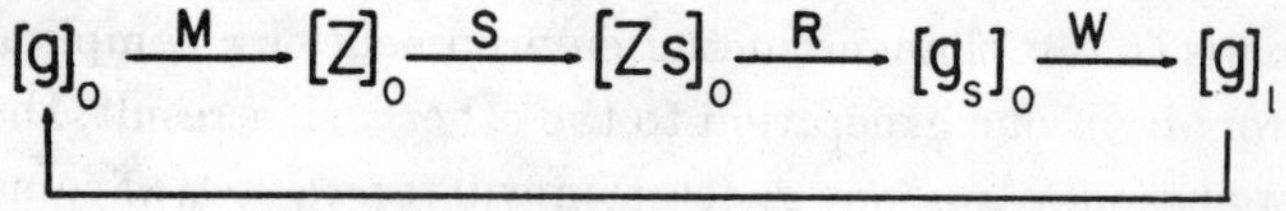

$$[g]_0 \xrightarrow{M} [Z]_0 \xrightarrow{S} [Zs]_0 \xrightarrow{R} [g_s]_0 \xrightarrow{W} [g]_1$$

FIGURE 1. The genetic transformation, T, broken up into its components during a single generation.

Figure 1 illustrates the basic model of genetics, which we use in one way or another in our investigations. You can think of the process that occurs in a genetical population as a transformation of a vector with 2^L elements from one generation to another. You begin with some vector that describes the population, it undergoes a series of genetical operations and comes out in the next generation with slightly different values. If the matrix describing the transformation is constant from generation to generation, then we have simple, well known ways of predicting what g_n will be from the higher powers of the matrix of transformation. We start with a vector of gametic frequencies, and that is turned into a matrix of zygotic frequencies by random mating. The actual operation is the multiplication of the vector by its transpose. This matrix, then, tells you the frequency in the population of each possible combination that has occurred in mating. The next thing that happens is that selection occurs. That is to say, the relative probabilities of survival and reproduction are taken into account. This produces a transformed zygotic matrix in which every element is multiplied by its W-value. The elements in the Z matrix add up to unity, while the elements in the transformed matrix no longer do so. The sum of the transformed matrix elements is $\bar{W}$. The selected individuals will now produce eggs in the next generation; that is to say, they are going to produce a new vector of gametes. This is done by the laws of Mendel, summarized in the transformation R. R includes Mendel's laws, and the phenomenon of recombination, in which one gene string recombines with another one. We know the exact rules of recombination and these rules are used to produce from the zygotic matrix a vector. Each element in the vector has a contribution from certain terms in the zygotic matrix. An algorithm can be constructed which produces the new gametic array from the zygotes. This new gametic array now has an array of new frequencies which no longer add up to one, because of selection. Each is divided by the normalizing factor $\bar{W}$, to produce a new vector g.

This continues generation after generation, and so, beginning with g_0, g_n is produced by applying this transformation n times. To short-circuit the process and go directly to the equilibrium condition, we find the g_0 such that the transformation turns out to be the identity transformation. This in turn involves the solution of a large set of simultaneous nonlinear algebraic equations. In easy cases, they can be solved by inspection. In most cases, only numerical solutions are possible.

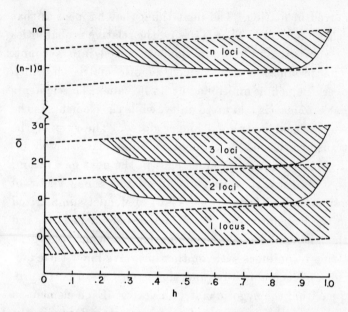

FIGURE 2. A generalization of the requirements for stability of the quadratic optimum mode. The ordinate is the value of the optimum $\bar{O}$ scaled in units of gene effect, a. The abscissa is the dominance h. The shaded areas are the regions of stability for successively larger numbers of loci. The dashed lines enclose the region of necessary stability.

Figure 2 illustrates the results of such computations for a model of selection in favor of an intermediate value of a character controlled by many genes. The shaded regions show the regions of parameter couplets $(\bar{O}, h)$ that tend to stable equilibrium of allelic frequencies. The exact meaning of $\bar{O}$ and h are unimportant in this context. We see a series of regions each corresponding to a different number of loci segregating. Each is stable. There are successive regions of stability for 2, 3, 4, 5, 10 loci segregating. Suppose there is an organism in which this character is controlled by 50 genes. All 50 of these genes can be kept segregated; that is, the system is strictly inside the hypercube of dimensionality 50. But if by any chance one of the genes is fixed at a frequency of 0 or 1, that is to say, if the system comes out to the face of the hypercube, then there is a new stable equilibrium configuration now strictly inside the $(n-1)$-dimensional hypercube. There is another stable point farther down, and another one farther down, and so on. Each one of these is a stable equilibrium and the stability of these points gets greater and greater as the dimensionality gets smaller. That is to say, a small perturbation will move the system from an n- to an $(n-1)$-dimensional stable state, a slightly larger one is required to move from $n-1$ to $n-2$, etc. The location of the stable point, when the dimensionality is very high, is very close to one vertex, but as you move the dimensionality down, the point moves farther and farther away from the vertex in the next lower dimensionality. Figure 2, then, is a case where there exist multiple stable configurations for a population, each one containing a different number of genes controlling the character, and each one successively more difficult to get out of, or successively easier to fall into as you go down. Any given population may occupy any one of these states, each with its own response to perturbation.

Table I is an example of the peculiarities of genetic potential surfaces. It shows the effect of changing one parameter only on the stability of an equilibrium. The one parameter is the frequency with which recom-

TABLE I

R	g_{00}	g_{01}	g_{10}	g_{11}	p	r	D'	$\bar{W}$
.00	.50000	0	0	.50000	.50000	.50000	1.00000	.95000
.01	.49667	.00333	.00333	.49667	.50000	.50000	.98658	.94000
.02	.49324	.00676	.00676	.49324	.50000	.50000	.97297	.93000
.03	.48979	.01021	.01021	.48979	.50000	.50000	.95916	.92000
.04	.48629	.01371	.01371	.48629	.50000	.50000	.94516	.91000
.06	.47913	.02087	.02087	.47913	.50000	.50000	.91651	.89000
.08	.47174	.02826	.02826	.47174	.50000	.50000	.88694	.87000
.10	.46409	.03591	.03591	.46409	.50000	.50000	.85636	.85000
.10 to .375				no stable equilibrium of gene frequencies				
.375 to .50	.25000	.25000	.25000	.25000	.50000	.50000	0	.57500

binations occur between the chromosomes. Provided recombination is relatively rare, the equilibrium frequencies of four gametic types that are in the population do not change very much. As the frequency of recombination increases, the frequency of types 00 and 11 drops slightly, the frequency of 01 and 10 increases slightly, but there isn't much change. Then suddenly, when the value of the recombination parameter exceeds exactly .10, the whole system lapses into a situation in which there are no stable equilibria at all. That is to say that everything will evolve to 100% 00 or 100% 11. There are no intermediate, nontrivial stable equilibria. Then, as the value of the parameter increases to .375, there suddenly occurs a new stable equilibrium point with equal frequencies of all four gametic types.

This is meant to demonstrate how complex these response surfaces can be, and how there can be the complete disappearance of stable equilibria with just a small change in a critical parameter.

Figure 3 illustrates another example of how changes in parameters can alter drastically and in a qualitative sense the genetics of a population. Here, instead of concerning ourselves with equilibrium, we are concerned with the path that the frequencies of the different genes in the population take over time. If the value R, the frequency of recombination of genes, is large, the particular selective process which is going on, which is also favoring an intermediate optimum, results in two of the genes becoming fixed at 100% of one allele, and two other genes becoming fixed at 0% of that allele. The fifth locus slowly goes to 100% after many, many generations, so that there is no intermediate stable

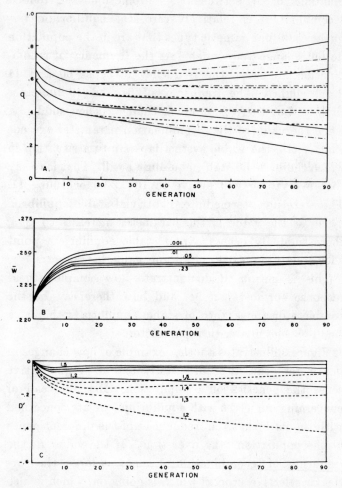

FIGURE 3. Changes occurring over time in a genetical system in which five loci are segregating. Top, changes in gene frequency (q) at the five loci for large recombination (solid lines) and small recombination (dashed and broken lines). Middle, changes in the mean fitness, $\overline{W}$, for different values of recombination. Bottom, changes in the correlations between loci for large recombination (solid lines) and little recombination (broken lines) for different pairs of genes indicated by the couplets of numbers.

equilibrium of allele frequencies. In other words, the population degenerates going to one of the surfaces of the hypertetrahedron. If, on the other hand, there is very little recombination, say 1%, then the loci form what is shown by the dashed lines, a sort of quasi-equilibrium situation. The population slowly goes to fixation but at a rate that is so slow that it is virtually impossible to measure it over several hundred generations. This means that in any real population, there is essentially a stable equilibrium, although not mathematically stable.

This brings up a last point, and that is when the biologist talks about stability, he doesn't really mean it. No biologist really thinks that any process in biology is 100% stable. What he means is that it is stable to all intents and purposes; will anybody live long enough to see it degenerate? It is important, very important, that we find an exact and useful formulation of the concept of the "degree of stability" of a biological process, rather than the more usual all or none concept.

Studies on Biological Clocks: A Model for the Circadian Rhythms of Nocturnal Organisms

Theodosios Pavlidis

I. Introduction. It has been established in recent years that many organisms present a periodic activity regulated by an innate oscillator whose period is approximately 24 hours. The adjective "circadian" (i.e. about a day) is commonly used in reference to such phenomena. It is beyond the scope of this paper to give even a rough review of the extensive work done in this field. The reader who is not familiar with the subject may find helpful the volume "Circadian Clocks", edited by J. Aschoff, which includes many examples of work in this field up to 1965.

It has been found that this oscillator can be phase shifted by light or temperature stimuli and, of course, it can be entrained by such inputs when they are periodic. Also, there has been evidence that the oscillator has only one degree of freedom and therefore, its state can be described by one parameter, the so-called circadian time (abbreviation CT). It has been found that the phase shift caused by an external stimulus depends on, among other things, the circadian time at the start of the stimulus. A plot of the phase shift versus CT (while the other features of the stimulus are kept constant) is usually called a *response curve*. (For more details, see Aschoff [2], Pittendrigh [13], and Pavlidis [10].)

This paper deals with the development of a mathematical model for the circadian oscillator of nocturnal organisms. The reason for this limitation is that light has opposite effects, depending on whether the organism is nocturnal or diurnal. It shows a tendency to damp the oscillation in nocturnal organisms, while the lack of light has a similar effect on diurnal organisms. Aschoff's rule summarizes this phenomenon: "Increasing light intensity lengthens the period and reduces the fraction of the period when activity is observed in nocturnal organisms, while it has the opposite effect in diurnal organisms" [1]. Although it may be possible to extend the model to diurnal organisms by some relatively minor modifications, this has not been attempted in this paper. At this point one immediately has to decide how to classify the exemptions to Aschoff's rule. A reasonable decision is to call an organism "nocturnal" if light damps out its rhythm, even if the organism is actually diurnal. Then the model which will be developed will be valid for all nocturnal organisms if "nocturnal" is interpreted as above. Under this assumption, the drosophila (a favorite subject in experiments related to circadian rhythms) is classified as a "nocturnal" organism although it is actually diurnal [13]. Only the light effects are studied in this paper. Temperature effects are described elsewhere (Zimmerman, Pittendrigh and Pavlidis [15]). This restriction is justified because when the temperature is held constant the circadian rhythms seem to be temperature independent, i.e. the period of the oscillation and the response curves to light stimuli change little, if at all (Zimmerman [15]).

In a previous study of the eclosion rhythm of the drosophila pseudoobscura (Pavlidis [9], [10], Pittendrigh [13]), a model was developed which gave a fairly satisfactory simulation of the effects of light stimuli on the phase of the rhythm. The following basic assumptions were made:

(1) That the total system consists of a primary subsystem which is a selfsustained oscillator and which drives a secondary subsystem (not necessarily an oscillator), the last one directly responsible for the eclosion. A simple argument shows that the time of eclosion triggered by the driving system corresponds to the circadian time where the response curve moves from delays into advances.

(2) That the steady state phase shift occurs immediately or in a fraction of a cycle in the driving oscillator and that

the observed transients are due to the driven system. (For details see Pavlidis [10].)

Under these two assumptions one can visualize a very general mathematical model: A dynamical system which presents a limit cycle (with one degree of freedom) and

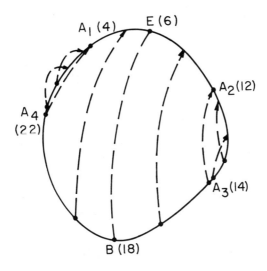

FIGURE 1. A topological model of the drosophila oscillator.

whose change of state because of a light stimulus of very short duration occurs almost instantaneously. The new state is also on the cycle or very close to it (so that no transients are observed). Then it is possible to map the response curve on a cycle as shown in Figure 1. Thus if light falls at CT = 16, the system moves immediately at CT = 8, which is obtained by subtracting from 16 the observed 8 hour delay for a 15 minute pulse occurring at CT = 16. Note that according to the model and the response curve, the light has no effect when it falls between CT = 4 and CT = 12 and also that almost all the other points of the response curve are mapped by light at this region.

The model is topological and the writing of specific equations is not necessary. Our description actually covers a class of mathematical models rather than a specific one.

Elsewhere it has been described how such a model was used to predict the effect of a combination of light pulses including entrainment by them (Ottensen [8], Pittendrigh [13], Pavlidis [10]).

After these encouraging results, an extension of the model was sought to describe its response to light stimuli of longer duration. The most natural assumption is that since the light has no effect in the zone of CT = 4–12, once the system is there it resumes its regular motion. If the light has been removed when the system arrives at CT = 12, then nothing else happens. If the light is still on then the system has to stay there until the removal of the light. This procedure makes it possible to determine the response curve to various light durations from the response curve to a 15 minute pulse. The results so obtained were in agreement with the experimental data. Entrainment of the rhythm by light stimuli of various durations was also simulated successfully (Pavlidis [10]).

The success of the topological model for the drosophila eclosion rhythm points to an important question:

Could one extend these results to other organisms? For example, how should the model be modified to account for the previously mentioned Aschoff's rule which deals with rhythms persistent in continuous light and not damped out like the eclosion rhythm?

To answer these questions we first have to look closer into the structure of the model: This will be the subject of the next section.

II. The development of a class of models for the circadian oscillator. The previously described topological model is very attractive because it makes no unnecessary assumptions about the nature of the system, but it has the disadvantage of not allowing any quantitative study unless it is of the all-or-none type. Therefore, before we proceed we have to choose a more restricted class of models. As such a class we choose oscillators with one degree of freedom. The general form of such an oscillator is given by the following pair of nonlinear differential equations

(1) $$dr/dt = f(r, s),$$
(2) $$ds/dt = g(r, s)$$

where r and s are the two state variables and f and g are to be chosen so that the system of equations (1) and (2) exhibits a limit cycle. For a physical system it is justified to assume that the range of values taken by r and s is not unbounded, but it is between certain limits. Moreover, from the previous model for the drosophila eclosion rhythm, one can conclude that the zone between CT 4 and CT 12 corresponds to the "saturation" of one of the state variables. We choose r to be this variable and without any loss of generality, we assume that it is always nonnegative, i.e. the "saturation level" is zero[1]. This choice is reasonable if we have in mind that r may not be an algebraic quantity, i.e. it may represent a physical variable where negative values have no meaning, as for example the concentration of a substance. On the same grounds we may choose to have s also limited to nonnegative values, although this restriction is not very essential to the subsequent development of the model.

In order that the system described by equations (1) and (2) has a limit cycle in the first quadrant of the $r - s$ plane, it is necessary that it has a singular point of Poincare index $+1$ in this quadrant (Minorsky [6, p. 79], Lefschetz [5, pp. 199–200]). Since the index of a point relative to the system of equations (1) and (2) is the same as for its first approximation (Lefschetz [5, p. 196]), the previous statement implies that the characteristic equation of the linearized system has two roots of the same sign. Let r_c, s_c be the coordinates of the critical point, i.e.

(3) $$f(r_c, s_c) = 0,$$
(4) $$g(r_c, s_c) = 0,$$

and also $r_c > 0$, $s_c > 0$. Also let f_r, f_s, g_r, g_s be the partial derivatives of $f(r, s)$ and $g(r, c)$ computed at the critical point, i.e.

(5) $$f_r = \partial f(s, r)/\partial r\big|_{r=r_c; s=s_c} \text{ etc.}$$

Then the linearized system is described by the following equations:

(6) $$dr/dt = f_r \cdot r + f_s \cdot s,$$
(7) $$ds/dt = g_r \cdot r + g_s \cdot s,$$

and the characteristic equation is

(8) $$\lambda^2 - (f_r + g_s)\lambda + f_r g_s - f_s g_r = 0.$$

The two roots will be of the same sign if

(9) $$f_r g_s - f_s g_r > 0.$$

Inequality (9) is a necessary (but not sufficient) condition for a limit cycle.

If the critical point is asymptotically stable, then there must be a region R around it so that if the point (r_0, s_0) belongs to R then the trajectory starting from it tends to (r_c, s_c) for $t \to \infty$. In terms of the model for the circadian oscillator this is undesirable. It would mean that once the state of the system is in R the oscillation would damp out and could be started again only by some outside stimulus. Since R would lie in the interior of the region enclosed by the limit cycle, the model should be designed in such a way as to exclude the possibility that a small light stimulus would bring the system into R. However, we have already seen from the first drosophila model a situation where the trajectories of the system cross R and one would expect them to terminate in R for small enough light intensities. Therefore, while a strong light stimulus would fail to damp out the oscillation, a weak one (of the same duration) would damp it. This is completely in contrast to the experimental findings and therefore, we conclude that the critical point should be unstable, i.e. the roots of the characteristic equation should have non-negative real parts. If they are zero, this would imply the existence of a center and the possibility of more than one periodic trajectory, depending on the initial conditions. This is also excluded, therefore the real parts should be positive, or in other words

(10) $$f_r + g_s > 0.$$

It is also preferable that the curves $f(r, s) = 0$ and

$g(r, s) = 0$ do not have more than one intersection, because that would imply the existence of another critical point in the plane and it would offer the possibility for large enough disturbances to move the system outside of the attraction region of the limit cycle (Nemytskii and Stepanov [7, p. 362]) and hence damp out the oscillation. (Such a situation has occurred in a previous model proposed by Wever [14].) The above conditions should be satisfied by any system which is a candidate for a model of the basic circadian oscillator.

We now want to choose one that will present a limit cycle which will include a segment of the axis $r = 0$ as part of its trajectory (in order to simulate the sector CT: 4–12). A very simple approach is to design the system in such a way so that the critical point is an unstable focus (i.e. we rule out now the possibility of an unstable node). Then all trajectories would eventually intersect the axis $r = 0$. There will exist one trajectory tangent to it, and the system will leave this axis on it (Point A) to meet it again at B (Figure 2). The curve $ACBA$ will be a limit cycle provided the motion along the s-axis is from right to left. It will be a stable limit cycle because any trajectory starting in the interior of the region enclosed by the curve $ACBA$ will eventually meet the segment BA and any one starting outside it will meet the s-axis ($r = 0$) and then the limit cycle at B. The necessary and sufficient conditions for a focus is that equation (8) has complex roots, hence its discriminant should be negative.

(11) $$(f_r + g_s)^2 - 4(f_r g_s - f_s g_r) < 0.$$

Note that if inequality (11) holds then so does inequality

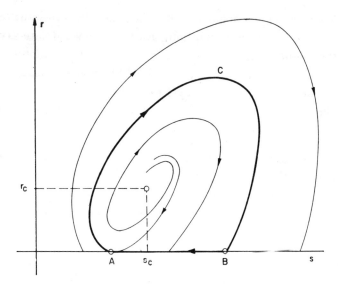

FIGURE 2. Phase plane plot of trajectories of the system of Eqs. (1) and (2).

(9), which then can be omitted from the list of necessary conditions for a limit cycle. Inequality (11) is simplified as

(12) $$(f_r - g_s)^2 + 4f_s g_r < 0.$$

For $r = 0$, equation (2) becomes

(13) $$ds/dt = g(0, s)$$

and the motion will be from right to left (i.e. s will be decreasing) if and only if

(14) $$g(0, s) < 0.$$

Conditions (10), (12) and (14) plus the requirement that the system of equations (3) and (4) have a unique solution in the first quadrant guarantee the existence of a unique stable limit cycle as shown in Figure 2.

The previous topological model can now be translated in terms of the new description. All that is needed is to say that light brings r immediately to zero level. A proper choice of the functions f and g should be made so that the response curve is reproduced exactly, and this is not too difficult. In § IV a specific example is given of a system satisfying the conditions derived in this section.

III. The effect of light.

In order to take into account the effects of variable light intensity we have to modify our previous description and devise a formal mechanism by which light affects r. The simplest possibility is to rewrite equation (1) as

$$(15) \qquad dr/dt = f(r, s) - K \cdot l$$

where K is a sensitivity coefficient and l represents the light intensity. If l is greater (in absolute value) than the maximum of $f(r, s)$ (with respect to both r and s on the limit cycle) divided by K, then dr/dt will be negative and r will start decreasing as soon as the light is turned on. The rate of decrease will be larger for larger l and in such a case r may be zero in a very short time. This corresponds to the drosophila case. For a given (small) pulse duration any increase of the light intensity beyond this level will not affect the shape of the response curve.

If the light stimulus is not very strong then the change in the value of r will not necessarily be such that $r = 0$, but r will have a value somewhere in between. This new value can be found approximately as follows. If the duration of the stimulus is very short, then the pulse can be approximated by a delta function of area L ($L = l \cdot \Delta t$, where Δt is the pulse duration) occurring at time t. Then, integrating equation (15) from $t - 0$ to $t + 0$ one obtains

$$r(t + 0) = r(t - 0) - K \cdot L.$$

After the removal of light the motion will resume according to the pattern shown in Figure 2 and the system will be back on the limit cycle after a time interval less than the period of the oscillation.

In such a case one may try to plot the response curve for various intensities. This was done by a computer simulation of a specific model described in § IV. The pulse duration was equivalent to 1 hour and the intensity was variable. The results are shown in Figure 3. The interesting finding is that the curves corresponding to lower light intensities in drosophila look very much like the response curves of mammals (DeCoursey [3], [4], Pittendrigh [12]). In view of the fact that mathematically only the product $K \cdot l$ is of importance for the shape of the curve, the following conjecture is suggested:

The same model can account for the response curves of the circadian rhythms of various nocturnal organisms provided that the sensitivity to light is different (i.e. the coefficient K is chosen to be different in each case).

If this is indeed true, then one may risk the conclusion that there is a common system in all organisms (or at least the nocturnals) which is responsible for the circadian oscillator. Such a system would most likely be at the subcellular level. However, it is very difficult to test this hypothesis any further now because of the lack of experimental data involving response curves. In spite of the large volume of experimental work in the field of circadian clocks, many interesting experiments have not been performed[2].

For example, there are no experimental results describing the effect of very low light intensities on the drosophila

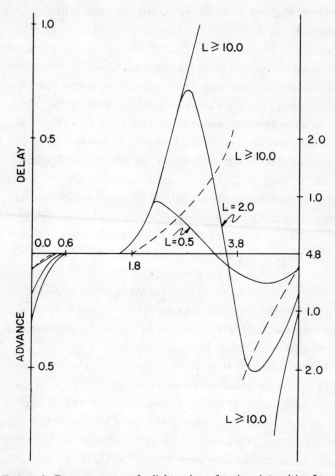

FIGURE 3. Response curves for light pulses of various intensities L and short duration (0.2). For $L \geq 10$, the response curves are identical. Because of the much larger phase shifts such a curve is replotted (broken line) in a smaller sclae (shown on the right-hand side of the graph).

eclosion rhythm. The only experimental evidence which corroborates our new results for the drosophila case is the response curve obtained for light flashes of very short duration (1/2000 sec.) (Pittendrigh [**11**]). From the model one expects that the response curve for such a stimulus will be similar to the one obtained for pulses of longer duration but of low intensity. Indeed, the 1/2000 sec. response curve looks very much like the curve for $L = 2.0$ of Figure 3. In order to check that further, response curves for stimuli of very short duration were also obtained as it is described in § IV. Equation (15) allows us to study the behavior of the system under constant illumination. The term $-Kl$ can be incorporated in f and this will have as a result movement of the curve $f(r, s) = 0$ in the direction shown in Figure 4a, b. One can see that the limit cycle (if it exists) will move towards lower r. For higher intensities the curve $f(r, s) = 0$ may move so far that it does not intersect the curve $g(r, s) = 0$ in the first quadrant. This means that the system will not have a singular point there, and therefore a limit cycle cannot exist.

Figure 4a shows the trajectories corresponding to low light intensity (or low sensitivity). A limit cycle exists and it is displaced only to the left. In this case the period is longer (see next section) and this satisfies Aschoff's rule for nocturnal organisms. If the threshold of activity is denoted by s_0 then one also sees that the proportion of the cycle where activity is present decreases and in certain cases no activity at all will be observed if the cycle lies to the left of s_0.

Figure 4b shows a case where no limit cycle can exist. This corresponds to high light intensities (or high sensitivity).

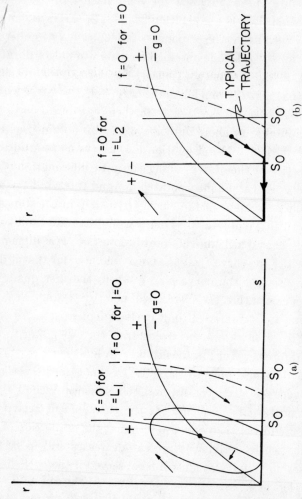

FIGURE 4. Change of the phase plane portrait under the influence of light. $L_1 < L_2$.

The model will present entrainment by periodic light stimuli but the study of this phenomenon will be the subject of a future communication. It must be mentioned that since for high light intensities this model reduces to the previous drosophila model, the mechanism of entrainment will be of interest only for weak light intensities.

A possible generalization of our description of light effects is to consider the possibility that the coefficient K is not constant but depends on the light intensity. In this way we are led to consider an adaptive system if K takes small values for high light intensities and large values for low light intensities. It is reasonable to assume that the change in the values of K is not instantaneous but it shows a time lag with respect to the change of the light intensity. This can be described in most simple terms by a differential equation of the following form:

$$\tau \, dK/dt + K = G(l) \tag{16}$$

where τ is a time constant and $G(l)$ is a decreasing function of l.

This feature will be very prominent when the system is exposed to light for periods much longer than τ and it offers an explanation for the behavior of the drosophila eclosion rhythm. It is known that although the rhythm is damped in continuous light, eclosion continues in irregular intervals. If we assume the existence of the adaptive mechanism, the oscillator is damped only temporarily, but as soon as K reaches a low enough value the oscillation starts again. Since τ may have different values for different individuals, one expects that the synchrony among various oscillators is lost and the apparent loss of rhythmicity is observed.

IV. Results of computer simulation. In this section we will describe some specific forms of the model developed in §§ II and III. We first notice that although the system described by equations (1) and (2) is nonlinear and as such could exhibit a limit cycle, this is not very pertinent to our development. The essential nonlinearity of the model is the condition $r \geq 0$. Thus the system would have a stable limit cycle even if equations (1) and (2) were linear. However, in order to have greater freedom in our effort to simulate the experimental results, a nonlinear system was chosen for specific study and this is described by equations (17) and (18) below.

(17) $\qquad dr/dt = r + d - c \cdot s - b \cdot s^2 - K \cdot l,$

(18) $\qquad\qquad ds/dt = r - as \quad (r \geq 0).$

All the coefficients are assumed to be positive.

Conditions (10) and (12) for this system are equivalent to

(19) $\qquad\qquad 1 - a > 0,$

(20) $\qquad (1 - a)^2 < 4\sqrt{(c-a)^2 + 4bd},$

while (14) is obviously satisfied. We also have:

(21) $\qquad r_c = \dfrac{a}{2b}(a - c + \sqrt{(c-a)^2 + 4ba}),$

(22) $\qquad s_c = \dfrac{1}{2b}(a - c + \sqrt{(c-a)^2 + 4bd}),$

and

(23) $\qquad s^* = \dfrac{1}{2b}(-c + \sqrt{c^2 - 4bd}),$

where s^* is the point of contact denoted by A in Figure 2.

A set of numerical values satisfying conditions (19) and (20) and giving positive r_c, s_c, s^* is

$$a = 0.5, \quad b = 1.0, \quad c = 0.6, \quad d = 1.0.$$

Such a system was simulated on a digital computer (IBM 7094) and its response to various light stimuli was studied.

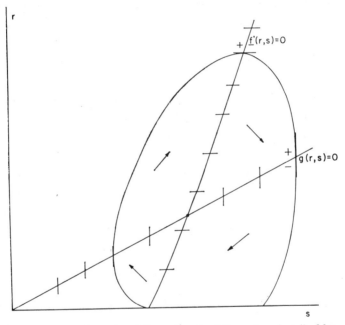

FIGURE 5. Phase plane plot of the unit cycle of the system described by Eqs. (1) and (2).

In order to test the theory of light adaptation, two models were simulated. One was with $K = 1 = \text{const.}$ and the other with K variable according to the differential equation

(24) $$dK/dt + K = 1/(1 + l).$$

The effect of such a mechanism becomes apparent only for light stimuli of high intensity and long duration. Therefore response curves for strong light and duration up to about half the period should not differ too much one from the other. This was verified by the computer simulation.

The computer simulation of Equations (17) and (18) gave a limit cycle shown in Figure 5 with period 4.77 (i.e. one unit equals approximately one fifth of 24 hours).

Subsequently the following experiments were simulated.

(1) Response curves for light pulses of duration 0.2 (i.e. equivalent to 1 hour) and variable intensity 0.5–40.0. The results are shown in Figure 1 and they were discussed in § III.

(2) Response curves for light pulses of high intensity (5.0) and variable duration, 0.8–2.0 (i.e. 4–10 hours). The results are shown in Figure 6b and they are in good agreement with the experimentally obtained response curves for drosophila. (Identical results were obtained when an adaptive system was simulated.)

(3) Response curves for light pulses of low intensity (0.5) and variable duration, 0.8–2.0 (4–10 hours). The results are shown in Figure 6a. These curves present no parallel with the known experimental data of similar situations, namely the response curves for luminescence of the Gonyaulax (3h. 1400 f.c.), leaf movement of phaseolus (3h. 10–150 f.c.) and petal movement of Kalanchoe (2h. red light) (Pittendrigh [12]).

Before attempting to adjust the model to simulate these data one would like to see more response curves obtained experimentally. To the author's knowledge there has never been derived a response curve for the same organism for various light intensities.

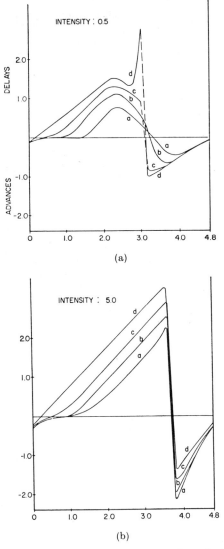

FIGURE 6. Response curves for variable duration. 6a is for low light intensity (or low sensitivity, e.g. mammals) and 6b is for high light intensity (or high sensitivity, e.g. drosophilia). In both graphs the durations are as follows: a: $0.8(=4h)$, b: $1.2(=6h)$, c: $1.6(=8h)$, d: $2.0(=10h)$.

(4) Behavior of the model under constant light for various light intensities. The results are shown in Figure 7 and follow Aschoff's rule, i.e. that for increasing light

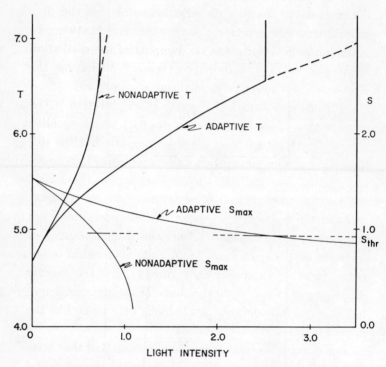

FIGURE 7. Change of the period as a function of the light intensity (Aschoff's rule) for adaptive and non-adaptive systems.

intensity the period of the oscillation lengthens. Since the activity starts when $s = s_{max}$, it may happen that although the system presents a limit cycle, no overt oscillation exists. In the graph, solid line indicates "overt period" and broken line indicates "hidden period". It is seen that for a nonadaptive system the oscillation is damped for a very low light intensity of about 1.0. In view of the shape of the response curve for such intensity (in Figure 3 it should be between the curve for $l = 0.5$ and $l = 2.0$) one can conclude that for the mammals this corresponds to average illumination while for the drosophila this corresponds to very weak illumination. In mammals it is certain that no damping of the oscillation occurs in average illumination levels; therefore, the existence of an adaptive mechanism is supported.

(5) Response curves for very short duration. The smallest step which could be used was 0.01, which equalled the size of the steps used for the integration of the differential equations in the digital computer. (This corresponds to 3 min., which is much longer than the 1/2000 sec. used in the experiment mentioned in § III.) Figure 8 shows the response curves for various intensities. For comparison, it is mentioned that the response curve of duration 0.05 (i.e. 15 min.) for $L = 20$ is identical to the curve for $L \geq 10$ shown in Figure 4. For the shorter duration even for $L = 20$ the shape of the response curve has changed considerably and it looks very much like the experimental curve of 0.5 msec.

At this point one may raise the objection that the "time scale" of the model is different from the experimental reality, since if it were possible to simulate on the computer 0.5 msec. response curve the phase shifts obtained would have been much smaller that the experimental ones. This disagreement is only partially true. The 0.5 msec. response curve was derived under much stronger light intensities (Pittendrigh [11]) than the 15 min. response curve. Hence although $l = 20$ seems an intensity corresponding to the one used for the longer duration response

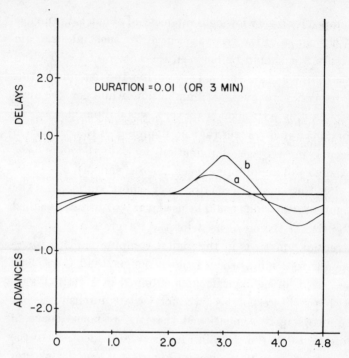

FIGURE 8. Response curves for light pulses of very short duration. Intensity is $l = 20$ (a) and $l = 10$ (b).

curves, a much higher value might be corresponding to the intensity used for the 0.5 msec. response curve.

V. Conclusions. In this paper a general model was described which seems to simulate a number of experimental data concerning the circadian oscillations of nocturnal animals. In particular, the model simulates Aschoff's rule and the response curves for various light intensities and durations. The most important conclusion is that the same mathematical model can simulate the behavior of more than one organism by changing only its sensitivity to light. This points out the possibility that a common circadian oscillator may exist at the subcellular level in all organisms.

VI. Acknowledgment.

The author wishes to express his thanks to Professor C. S. Pittendrigh of the Department of Biology of Princeton University for his many helpful comments and discussions during the development of the model which is described in this paper.

The computer facilities used for the simulation described in § IV were supported in part by National Science Foundation Grant NSF-GP-579.

Footnotes

[1] Indeed, assume that the system is such that r is always less or equal to some other level R and is represented by equations (1) and (2). Define a new variable $r_1 = R - r$. This will always be nonnegative. Then (1) and (2) can be rewritten as

(1') $\qquad dr_1/dt = -f(R - r_1, s) \triangleq f_1(r_1, s)$

(2') $\qquad ds/dt = g(R - r_1, s) \triangleq g_1(r_1, s)$

and the new system of (1') and (2') is of the same general form as the system represented by (1) and (2).

[2] This is mainly due to the long time required for experiments involving the circadian oscillator. The derivation of the response curve for a single stimulus (i.e. of fixed duration and intensity) may easily take six months. On the other hand a similar derivation for a model simulated on a high speed computer may take only six seconds!

[3] *C.C.* refers to the volume *Circadian Clocks* edited by J. Aschoff and published by the North-Holland Publishing Company in Amsterdam, 1965.

[4] CSHSQB below refers to the Proceedings of the Cold Spring Harbor Symposium on Quantitative Biology, 1960.

References

1. J. Aschoff, CSHSQB[4] **25** (1960), 11.
2. ———— "Response curves in circadian periodicity" in *C.C.*,[3] pp. 95–111.
3. P. J. De Coursey, *Phase control of activity in a rodent*, CSHSQB[4] **25** (1960), 49–55.
4. ———— *Daily light sensitivity rhythm in a rodent*, Science **131** (1960), 33–39.

5. S. Lefschetz, *Differential equations: Geometric theory*, Interscience, New York, 1963.
6. N. Minorski, *Nonlinear oscillations*, Van Nostrand, Princeton, N.J., 1962.
7. V. V. Nemytskii and V. V. Stepanov, *Qualitative theory of differential equations*, Princeton Univ. Press, Princeton, N.J., 1966.
8. E. Ottensen, *Analytical studies on a model for the entrainment of circadian rhythms*, A.B. Thesis, Princeton University, Princeton, N.J., 1965.
9. T. Pavlidis, *An analytical model for the drosophila eclosion rhythm*, Proc. 18th Annual Conference on Engineering in Medicine and Biology, 1965, p. 183.
10. ———, *A mathematical model for the light affected system in the drosophila eclosion rhythm*, Bull. Math. Biophysics **29** (1967), 291–310.
11. C. S. Pittendrigh, *Circadian rhythms and the circadian organization of living systems*, CSHSQB[4] **25** (1960), 169–184.
12. ———, "On the mechanism of the entrainment of a circadian rhythm" in *C.C.*[3] 1965, pp. 277–297.
13. ———, *The circadian oscillation in drosophila pseudoobscura pupae: A model for the photoperiodic clock*, Z für Pflenzenphysiologie **54** (1966), 275–297.
14. R. Wever, "A mathematical model for circadian rhythms" in *C.C.*[3], 1965, pp. 47–63.
15. W. F. Zimmerman, C. S. Pittendrigh and T. Pavlidis, *Temperature compensation of the driving oscillation in drosophila pseudoobscura and its entrainment by temperature cycles*, J. Insect Physiology (in press).

PRINCETON UNIVERSITY

Author Index

Roman numbers refer to pages on which a reference is made to a work of an author.

Italic numbers refer to pages on which a complete reference to a work by the author is given.

Boldface numbers indicate the first page of an article in this volume.

Aschoff, J., 89, *111*

Corbett, H. S., 30, *38*

De Coursey, P. J., 99, *111*
Doob, J. L., *45*

Elton, C., *14, 38*

Fisher, R. A., *38*

Gause, G. F., *13*
Gerstenhaber, M., 2
de Groot, S. R., *52*

Hairston, *14*
Huffaker, C. B., 53, *61*
Hutchinson, G. E., *14*

Kac, M., *14*
Kerner, E. H., 2, 10, *14*, 27, 31, *38, 45*
Khinchin, A. I., 9, *14*

Lefschetz, S., 94, *112*
Leigh, E. G., Jr., **1**
Levins, 62, 69
Lewontin, R. C., **62**
Liapunov, A. A., *52*

Lindeman, R. L., *14*
Lotka, A. J., 3, 5, *13*
Luce, 70
MacArthur, R. H., 2, *14,* 57, 62, 67
MacLulich, D. A., 56, *61*
Mazur, P., *52*
Minorsky, N., 94, *112*
Nelson, E., 12, *14*
Nemytskii, V. V., 96, *112*
Ottensen, E., 92, *112*
Pavlidis, T., **88**, 89, 90, 91, 92, *112*
Pittendrigh, C. S., 89, 90, 92, 99, 101, 106, 109, 111, *112*

Raiffa, 70
Slobodkin, L. B., *13, 14*
Smith, *14*
Stepanov, V. V., 96, *112*
Thom, R., 79
Volterra, V., 3, 5, *13, 38*
Wever, R., 96, *112*
Wiener, N., 13, *14, 45*
Williams, C. B., 30, *38*
Zimmerman, W. F., 90, *112*

Subject Index

Arctic foxes, 43
Aschoff's rule, 89, 92, 108, 110
autocorrelation, 45

biological association, 3
biomass, 10, 36
Boltzmann, Maxwell-, distribution, 39
Brownian motion, 12

Canadian lynx, 55
canonical average, 23
character displacement, 7
chromosomes, 75
circadian, 88
    rhythm, response curves of, 99
    time, 89
coefficient of interaction, 17, 19

community, 3, 7, 10, 35
competition, 5, 48
competitive exclusion, principle of, 6, 7
crashes, 11
cross-correlations, 47
crowding effect, 46, 55
cyclic fluctuation, 7, 15

distribution, of animals, 65
drosophila pseudoobscura eclosion rhythm, 90, 101
dynamical
    equation, 42
    law, 4

ecology, 3
elementary interactions, 15

equilibrium stochastic theory of island populations, 67
ergodic theorem, 22
evolution as an optimizing process, 69
evolutionary genetics, 72
experimental example, 5
extinction curves, 68

fitness, 76
fluctuation, 35
food web, 37
   diagram, 19

gametic frequency, vector of, 80
Gaussian distribution, 12, 25, 26, 39
   Markov process, 39, 48
genetic
   potential function, 76
   systems, complex, 62
genetical
   configurations, 72
   operations, 80

immigration curve, 68
insect outbreak, 38
insular zoo-geography, 68
intermediate optimum, 85

Lagrange multipliers, 36
lemmings, 10
linear approximations, 16, 31, 49
   utility functional, 71
logistic growth-law, 8
logistically regulated population, 42

matrix of zygotic frequencies, 80
Maxwell-Boltzmann distribution, 39
Mendelism, 73
microcanonical measure, 28

neighborhood stability, 78
optimal behavior of populations, 69
optimality, 69
oscillation, 9, 53
   predator-prey, 53

Poincare recurrence theorem, 24
population
   biology, 63, 64
   crashes, 36
   curve, 32
   explosion, 10
   genetics, 73
potential
   surface, 65
   well, 65
predation, 48
predator, 8, 15, 53
   -prey oscillation, 53
prey, 6, 8, 15, 53
primary producers, 19
productivity, 10, 35

quasi-equilibrium, 87

recombination, 81
response curve, 89, 106, 109, 110
   of the circadian rhythms, 99
   of various intensities, 99

selective process, 85
specialization, 38
stability, 10, 30, 35, 48
standing crop, 10
state-vector, 3
statistical
   formalism, 9
   mechanics, 9
stochastic
   equation, 41, 51
   logistic, 43
structural instabilities, 79

time lag, 18
transitivity, principle of, 70
trees, 7
trophic levels, 11
turnover rate, 37, 38

varying hare, 55
vector of gametic frequencies, 80

yeast, 5, 7

zygotic frequency, matrix of, 80